LANDS AND LEAD MINERS

A HISTORY OF BRASSINGTON
IN DERBYSHIRE

LANDS AND LEAD MINERS

A HISTORY OF BRASSINGTON IN DERBYSHIRE

RON SLACK

First printed by the author
This edition by Tempus Publishing 2007

Reprinted 2017 by
The History Press
The Mill, Brimscombe Port
Stroud, Gloucestershire, GL5 2QG
www.thehistorypress.co.uk

British Library Cataloguing in Publication Data.
A catalogue record for this book is available from the British Library.

ISBN 978 0 7524 4497 0

Typesetting and origination by NPI Media Group
Printed and bound in Great Britain by TJ International Ltd, Padstow, Cornwall

CONTENTS

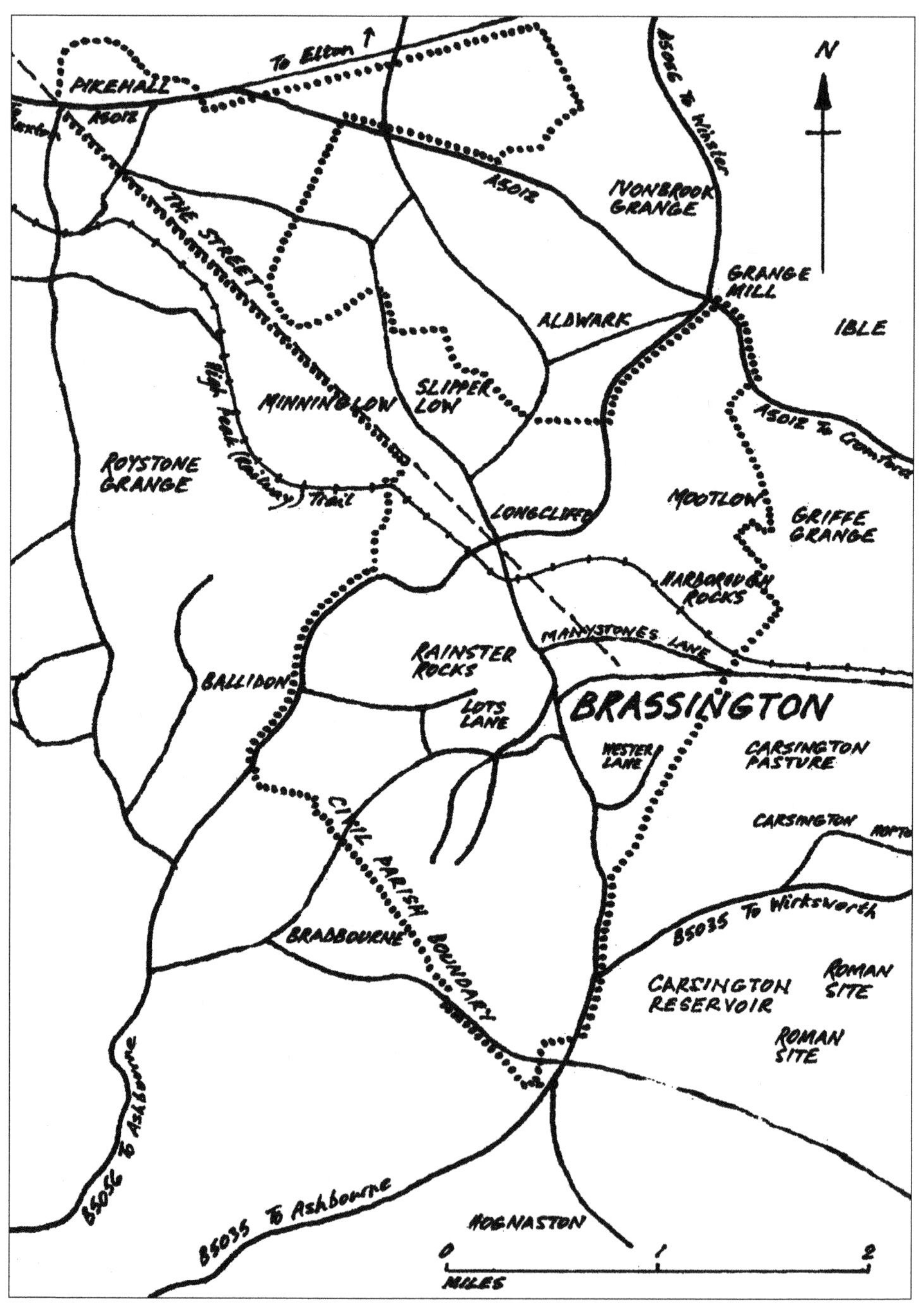

Map of the area.

FOREWORD

In the summer of 1976, on the recommendation of a keen rambler of our acquaintance, my wife and our family of four young children walked to Brassington on the High Peak Trail, the former Cromford and High Peak Railway, leaving the car at what had been the Minninglow Station. Brassington was my ancestors' village though, apart from attending family funerals, this was my first visit. In the same year the last of my grandfather's generation died and I was subsequently given pages from a family Bible, which recorded births, marriages and deaths from 1835 to his own birth in 1889, when his father, my great-grandfather, closed the record in Biblical fashion – 'Here endeth our increase of families.' I was later given my great-grandfather's portrait photograph, which since about 1880 had hung in the cottage and which I later found to have been in the family since the eighteenth century. In 1982 I started family-history research. This led to village history, lead mining and a biography of Sir John Gell of Hopton who, among greater claims to fame, controlled most of the seventeenth-century lead industry in the area.

The pleasure of research lies partly in detective work among old documents and printed books and articles. There has been an equal pleasure, for me, in discovering the present-day Brassington, talking to the people who live there and walking the fields and hills around it. Any history is an interpretation and this is my interpretation of what I have found out about the lives of the people who have lived in Brassington.

I have not translated sums of money into decimal, or weights and measures into metric, although archaic measurements such as oxgangs and more obscure terms such as bushels or pecks are explained. To give the decimal equivalent of a sum in the old coinage would make it even more difficult to appreciate than the vast changes in value have already made it. However, £-s-d are rapidly receding into history and it may be useful to note that a shilling (1s) translates as 5p, that there were twelve pennies (12d) in a shilling and that there were twenty shillings to a pound.

For their help I thank Roger Flindall, Lyn Willies, John Jones, the late Mr and Mrs Horrocks, Sam Bacon, the late Mary Walker and Doug Slack, Clayton Slack, Nigel Taft, Bethan Camacho, John Linnell Mike Rose, Edward Garner and Andy Wood. The late Harold Brittain supplied me with copies of documents, maps and old photographs, told me many things about the village from his own knowledge, and introduced me to other villagers, to whom I am also indebted. The 1967 *History* and its lead-mining supplement of 1971, produced in the village, have been invaluable reference sources. Rob Rowbotham drew the maps. I acknowledge with thanks the help of the staffs of the National Archives, Derbyshire Record Office, Derbyshire County Library, Sheffield City Archives, Lichfield Joint Record Office and the John Rylands Library, Manchester.

Photographs are reproduced by permission of Lynn Willies (p. 29), Clyde Surveys Ltd (p. 17), Derby Museum (p. 46), Michael Platt (p. 71), Derby City Library (p. 81), and Barry Robbins (p. 98).

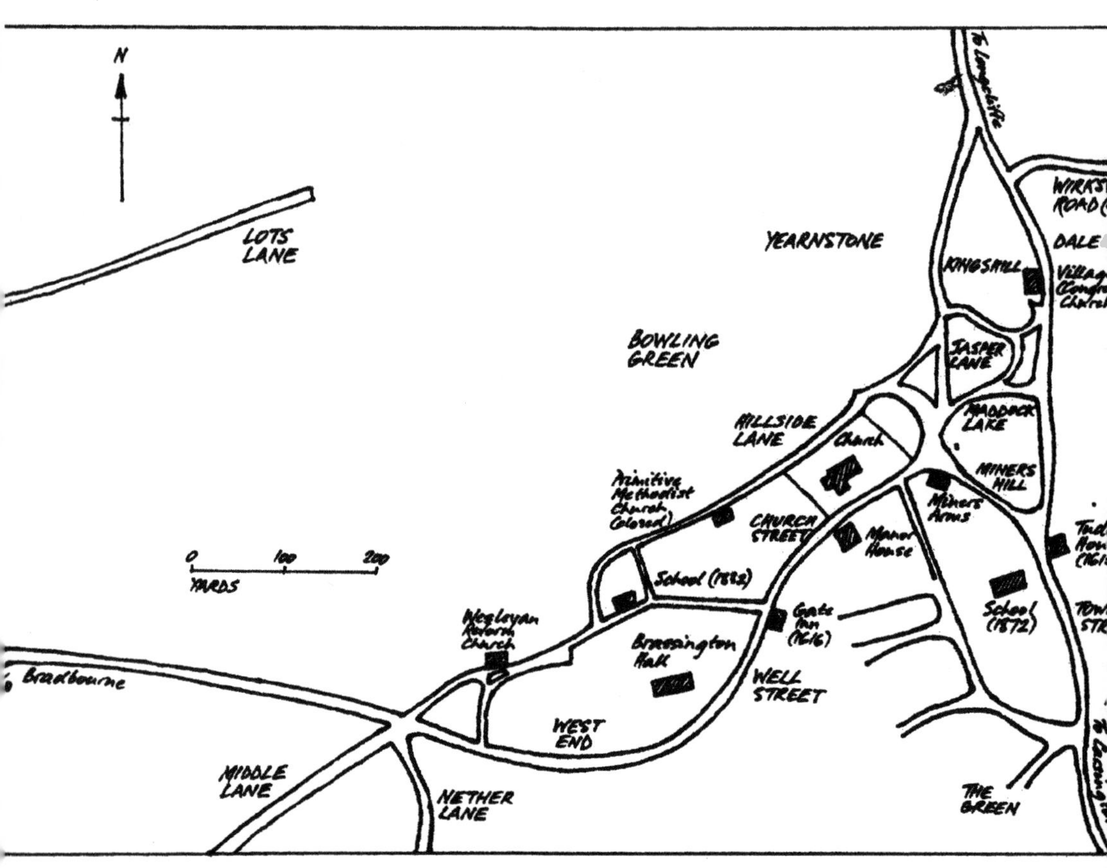

Map of the Village

CHAPTER 1

THE FIRST PEOPLE

THE BUILDINGS

One bitter winter's night in December 1680 a certain Michael Adams, Bachelor of Divinity, rector of Trenton in Yorkshire, was travelling on the road through Brassington which for centuries had been the main road from the south, when he succumbed to a fever and died. There's a brass memorial in Brassington church which, in Latin, links the fever raging within him with the blizzard raging outside. Sixty years earlier, the leading farmers of the village, composing a document to be presented to their landlord, the duchy of Lancaster, explaining why their rent should not be raised, decided to include the weather as one of the drawbacks of living in Brassington – 'the clymate is so could in winter.'[1] To the surveyor John Farey in 1817 it was 'so small and obscure a town'[2] and it still is off the beaten track, long bypassed by the main roads through the district. Yet Brassington is a good place to live, and always was. The weather has improved since the seventeenth century and modern settlers who come for the rough beauty of this limestone village in its rocky setting rarely find themselves stranded in snow drifts. What attracted the invaders who moved into the area in the sixth century was the shelter afforded by the narrow cleft in the limestone from the winds of the open plateau to the north, and the stream running down to level ground to the south. On the surrounding moorlands they found a treasure which supported the villagers for centuries – lead ore. This they mined in their own way, following what one eighteenth-century critic called 'Brasson Laws.'[3]

It was during the best times for the lead miners (from the seventeenth to the nineteenth century) that most of the houses were built. Viewed from its high western edge on Hillside Lane, Brassington is a huddle of limestone cottages, bunched at the northern, highest part of the valley in which the village lies, and climbing its western slope. Most of the oldest buildings are small. Two exceptions are the church, its Norman tower buttressed on the south to resist the steep hillside, and a large Jacobean house on the east of the village, now called the Tudor House. For much of its history this was a thriving inn on a busy road. There are twentieth-century houses at the south of the village, but most of the narrow, steep streets lead to houses that have been there for anything from 100 to 400 years.

PREHISTORY

Looking over the village toward Carsington Pasture, the rocky plateau is seen covered in the easily recognisable remains of hundreds of the small mines which were the livelihood of the people who built the cottages. The villagers' work took them into the rough landscape to the east, north and west of Brassington, and here the signs of their mining and farming are mingled with the remains of earlier settlements. Field boundaries built on Carsington Pasture before the Romans came are only partly hidden by the miners' shafts and hillocks. To the west, at Rainster Rocks, successive generations from earlier than the Christian era up to the nineteenth century have left signs of their work.

The village from the eastern slopes.

At a point where the ridges and furrows of medieval ploughland reached boulder walls built by the native British during the Roman occupation there is a deep trench, carved into the hillside by eighteenth-century miners who knew nothing of their British and Saxon predecessors. There is no way of knowing exactly when the English villeins cleared the scrub and ploughed up to the old settlements, but, standing below Rainster Rocks, it is very apparent that after the British had abandoned their settlement and its walls had sunk into the landscape, their successors extended their ploughland onto the ground they had left, no doubt turning up coins and weapons, tools and ornaments which amazed them. Later, after the new ploughland had itself been abandoned and put to sheep pasture, came the miners. The outline of the history of the village and its parish can be seen everywhere – nineteenth-century enclosure walls cross eighteenth-century mining rakes, medieval ridges and Romano-British settlements.

Brassington's name is Anglian. The Anglo-Saxons, arable farmers in their north-European homeland, settled in valleys when they colonised Britain, though they were probably not the first people to settle where this valley makes a north-south break in the limestone plateau. Their predecessors had also lived on the uplands. When the Roman legions marched from their fort at Little Chester, near Derby, to their base at Buxton, they marched on a road which took them past the future site of Carsington, round the north of the valley in which Brassington now lies and onward via Minninglow to Buxton and beyond. At Carsington their march took them out of the fertile, populated lowlands on to the edge of the Peak, where the rock was close to the surface, with few trees and only scattered native settlements. Their road crossed Carsington Pasture, through the rock outcrops and past the imposing monumental Harborough Rocks. There had been people living there, in the shelter of the rocks and in the caves among them, for centuries. The entire plateau on the west and east of Brassington's valley had been sparsely settled for at least three 3,000 years by the time the Romans marched across it.

From the time when the hunters of the Old Stone Age gave way to settled communities, there were farmers on the plateau. We have no idea what the earliest people called themselves or what others called them since neither they, presumably illiterate,

One of Brassington's narrow, winding side streets.

Terrace settlement site below Harborough Rocks.

nor anyone else, wrote anything about them. They did, however, leave evidence about themselves whenever they buried or cremated their dead. They buried vessels, tools, weapons and ornaments with the bodies of their dead neighbours. One of the earliest groups to have been found was buried to the north of Brassington, at Slipperlow. Here the antiquary Thomas Bateman dug into a mound in 1844 and uncovered a crouched skeleton in a rock grave.[4] In the same mound were three others, which had been disturbed at some time in the four millennia since they had been buried. These people had been buried with a pottery vessel which archaeologists call a beaker, and because similar communities in Britain always included at least one beaker in their burials they have been given the name Beaker Folk. They were familiar with the use of metal tools though they were not metal-workers themselves and continued to use flint tools similar to the ones used by the earlier Neolithic people.

The Beaker Folk on Brassington Moor 4,000 years ago were living during the latest stage in a long occupation which had seen the same sort of changes in technical development, in fashions in pots and pans, in immigration of new groups as have happened throughout history and are happening now. Their predecessors of the Neolithic or New Stone Age left the remains of a settlement at Minninglow[5] which persisted through into the time when people were using bronze tools.

Minninglow was a burial ground during the Beaker and Bronze Ages and, though the earth mound which covered them has not been replaced since Bateman dug it away 150 years ago, the great rectangular capstones of the chamber tombs are still there, resting on the limestone slabs which have supported them for 4,000 years. The Beaker graves at Slipperlow demonstrate the continuity between the different communities. Above the Beaker burials Bateman found a cremation, accompanied by two flint axe heads and two other worked flints, characteristic of the early Bronze Age. Burial grounds seem to have retained their sanctity for many centuries – another Bronze Age burial at Galleylow[6] had the skeleton of an Anglian warrior buried on top of it two thousand

Chamber tomb capstones at Minninglow. Minninglow can be seen from Beeley Moor, ten miles away, to the north of the Derwent Valley.

years later. Other burials include a grave on top of Harborough Rocks excavated in 1889[7] and a cremation at Mootlow discovered by Bateman in 1844.[8] The Harborough grave had sixteen skeletons, six of them in an undisturbed crouching posture, like the one at Slipperlow. The grave was also similar to the one at Slipperlow in having been built as a cairn. It contained three leaf-shaped arrowheads. The Mootlow cremation dates from the Bronze Age, about 1500 BC, and had an urn with a collar, and a bronze razor. In 1890 two young brothers, Edmund and Isaac Rains, rescued a sheep from a cave south of Longcliffe and discovered another grave, this one containing the crouched skeleton of a woman who had died during the early Bronze Age.[9]

These people of the plateau, adapting to changing techniques in farming, its tools and methods, to changes in fashions of pottery and ornament, to influxes of newcomers bringing new practices and beliefs, made a living for themselves for several thousands of years. They survived, though one fact which emerges from the Neolithic burials is that they did not live to what we would think of as an old age, and a disproportionate number of the skeletons are children's. The Neolithic graves were the last resting places not of individuals but of groups, and the people seem to have buried their dead in the same grave at a number of times, and in some cases to have removed skeletons from one grave and reburied them in another. The account of the Harborough excavation describes the remains of 'at least' sixteen people – some of the bones could not be sorted with any certainty into a definite number of skeletons as they had been intermingled when they were put there. This grave was of the type known as a 'passage grave'– it was constructed of boulders, with a chamber, and covered with a cairn of earth.

The people who built this tomb, and used it for many generations, lived at the same time as the people who raised the Nine Ladies circle of stones on Stanton Moor and the large circle at Arbor Low, near Youlgreave. The Harborough people may have lived in the large cave there, but are just as likely to have built themselves wooden houses – men who had the skill to fashion flint into arrowheads and axes were not likely to have found house building beyond them. Their field boundaries are still to be seen below the rocks, and the boundary walls on Carsington Pasture may have been their work. Similar walled fields cleared by their contemporaries have been found at sites all over the British Isles and the Harborough settlement, probably a few farms worked by inter-related families, must have been in touch with other communities since there is no flint near Brassington. Harborough is, in fact, close to the Portway, the route travellers were using at least as early as the Neolithic period. Their farms were sufficiently productive for them to be able to spare time for tomb and monument building, and while it is impossible to be sure of what the circles meant to them, their mental horizons clearly extended beyond the harsh business of making a living on the bleak plateau.

Change came slowly, one way of living overlapping the last one. The Harborough tomb's six crouched skeletons came from a community which shared enough earlier beliefs to use the ancient burial place, but had evolved a new way. They buried their dead once and left them undisturbed. These crouched skeletons are the remains of people from the same culture as the man at the bottom of the burial mound at Slipperlow. Communal graves gave way to individual ones – the Slipperlow man was buried alone in his rock grave – and people began to bury pottery with their dead. Their beakers were decorated and shaped something like a modern beer mug, often having a handle on one side. The Slipperlow grave's beaker dates it to the beginning of the time when flint tools and weapons were gradually superseded by bronze ones. Bronze was still rare and valuable and at first the new tools were imitated in flint, like the two axe heads and other tools in the cremation found above the grave at Slipperlow. The mound at Mootlow is another example of a cremation which took place above an existing grave – under the cremation, with its urn and the bronze razor, the only bronze object found on the plateau, Bateman found a crouched skeleton buried on a layer of rough limestone.

The people used other materials as well as flint and wood, before metal tools became common. A burial at Galleylow[10] had a piece of ivory, with six perforations, and a piece of stag's antler. The ivory may have been part of a necklace. The antler was probably a tool – they were used as picks. This grave consisted of two cairns, each surrounded by a ditch, when it was excavated in the eighteenth century, and contained several skeletons, including a child's, and a cremation. There was also a food vessel and a small cup. All these burials are the last resting places of ordinary people. There are no signs of wealth and none of the tombs are large or monumental. Through all the changes, the people of the plateau carried on with their lives, enduring the bleak winters, watching their crops grow in the summers and rearing their animals. They were not so isolated that they had no contact with others – the flint and pottery can only have come from other parts of the country, and indeed the beaker pots were of a pattern which was common throughout Europe, and may have come from the continent.

ROMANS

This constantly evolving pattern of tools, pots, weapons and customs continued, and during the period of about 800 years before the Romans came in the first century AD, there were people living at Harborough[11] who used iron tools. They also wore fine jewellery – a coral brooch and a bronze pin have been found there. They wove cloth, though perhaps not at Harborough since only spindles and weaving combs have been found, with no sign of loom weights. The people living at Harborough at this time, and their neighbours on the plateau, found themselves caught up in the effects of high politics, after the arrival of the Emperor Claudius' legions in AD 43. The plateau was on the southern edge of the tribal kingdom of the Brigantes, whose heartland was in Yorkshire, and was bordered on the south by the territory of the Brigantes' neighbours, the Coritani. The queen of the Brigantes, Cartimandua, was happy to keep on good terms with the Romans, who for a number of years remained in the south of the island, now officially identified in their histories as Britannia. The situation changed irrevocably for the people of the plateau when Cartimandua was ousted by an anti-Roman faction led by her husband, Venutius, and Brigantia became an enemy territory which the Romans had to conquer. They advanced from the south some time after the year 69 and defeated Venutius in 74.

The Romans were interested first in establishing secure bases for the legions, with rapid communication between them. The road they built from Little Chester to their new camp at Buxton, known for centuries as 'the Street'[12], skirted Brassington's valley to the north and has been traced to within a few hundred yards of the village, between Manystones Lane and the former Wirksworth Road. Its name survives in a field name, Street (or Straight) Knowl, at Longcliffe. Once the legions had built their road they set the people to work. The prize for the Romans was lead ore. The plateau was rich in veins of lead and it is likely that the natives were already skilled at extracting it – one of the reasons why the emperors wanted to add this cold northern island to their territories was that it was known to be rich in minerals.

A lead-working site was set up near the Scow Brook in an area now innundated by the Carsington Reservoir[13], with a settlement nearby which included a villa[14] which may have been occupied by the manager of the enterprise. There is no direct evidence that the army set the local people to mining. The Roman occupiers, however, were soldiers and administrators, followed by traders and merchants – the work must have been done by the natives, especially as the Romans regarded mining as work fit only for slaves or barbarians. There is also no evidence for deep mining. No Roman shafts have been identified on Carsington Pasture and it is likely that at that time there was plenty of lead near enough to the surface to be extracted without the need to sink shafts. The evidence for the Roman mining industry is in the shape of 'pigs' or ingots of lead found in a number of places in Derbyshire, including the Brassington area. Most of the pigs are stamped with an inscription which includes the name Lutudarum. There is agreement among archaeologists and historians that Lutudarum was probably in the Matlock/Wirksworth area. It may have been the settlement at Carsington. Certainly one of the pigs was found on the Brassington/Carsington parish boundary and there are two very roughly cast pigs, whose origin is not in doubt. These were found at Carsington, in a shallow pit where they had been stacked sometime during the second half of the fourth century.[15] The pit had pottery fragments and traces of lead and charcoal and lead was obviously being worked nearby. It was about a mile from the remains of the villa.

The people who lived during this time, from the second century to the fifth, were the first in all the generations who had lived and died on the plateau to see irresistible power and enviable wealth. The army may have built its road over the plateau between Carsington and Brassington because it knew about the lead deposits there. It may have been lucky chance for the Romans that such extensive deposits proved to be on their line of march. Whichever is true, the lead was profitable. Won from the limestone on the plateau, the lead was probably smelted elsewhere, since the valley site at the Scow Brook seems unlikely to have been suitable for smelting. It seems possible that the rough ingots were melted on the Scow Brook site, recast as the Lutudarum pigs and distributed via the nearby Street. The income from this industry sustained a sizeable settlement for about 200 years.

The villa at Carsington had refinements that must have amazed the locals. While they knew only cave homes or huts, their master lived in a stone house with several rooms, roofed with sandstone slates. It was a rectangular building with a large central hall with two rooms to the north of it and three at the south. One room of the villa had under-floor heating and there were mosaic floors, their patterns formed from a combination of

Remains of settlement walls at Rainster Rocks.

orange chert tiles and yellowish gritstone ones, both from local quarries. Added later to the south-eastern corner of the building was a truly Roman extravagance, a bath house, with under-floor heating and a terracotta-tiled roof.[16] Either this villa or an earlier one on the same site had glazed windows, a feature which, like slated or tiled roofs, or central heating, or regular bathing, or solid roads, was to disappear for over 1,000 years when the Romans left. The large central living room had two fireplaces. The people of the villa used pottery dishes, bowls, jars and beakers from potteries in Derbyshire, Lincolnshire, Dorset and across the Channel in Gaul. This was not the house of a rich man by Roman standards, but it clearly belonged to a family of standing, with a comfortable standard of living which persisted for more than 200 years.

The British who mined the lead ore on Carsington Pasture or worked it at the Carsington settlement may have been from the settlement at Harborough or from the substantial ones at Roystone Grange,[17] near Ballidon, or Rainster Rocks.[18] The cave at Harborough continued in occupation during the Roman period. Excavations have uncovered spears and iron tools dating from then, Roman coins and a brooch with the image of the Roman goddess of wisdom, Minerva. The buildings at Roystone included an aisled timber building measuring twenty metres by ten which was later rebuilt in stone on a slightly smaller scale (fourteen by eight). The wooden building was there at the same time as the villa and smelter at Carsington, and the people who lived in it and farmed the fields nearby, whose boundary walls have been excavated, probably divided their time in the same way as their successors did until quite recently; between farming and mining.

At Rainster Rocks, where lines of boulders (orthostats) remain from some of the settlement walls, excavators have found bits and pieces which establish that people were living there during Roman times. There were fragments of pottery from the third century, brooches, a knife, a whetstone, a fragment of a quern used to grind corn, a furniture handle and, remarkably, a pair of bronze dividers. This measuring instrument has a knob with two decorated collars. There were coins of the third and fourth centuries, including one which was found by a Brassington villager and is still in the village, less than a mile from where it was lost 15,000 or 16,000 years ago. This was issued by one of the Caesars, who were subordinate emperors installed first by the emperor Diocletian to spread the burden of government. It has the youthful face of Constantine, eldest son of Constantine the Great, the emperor born in Britain and the one who made Christianity the state religion. The younger Constantine was made Caesar of the Western Empire, with his base in Gaul, in 317. Also found at Rainster were two pieces of smelted lead and a lump of lead ore. If these are the same age as the other finds, the settlement was a second site of lead working. The miner-farmers were exporting lead and wool for cash. Export trades need secure communication and the deterioration of the empire at the end of the fourth century, when Germanic pirates made the North Sea crossing a hazardous enterprise for traders carrying lead to the continent, damaged the communities at Roystone, Rainster and Carsington.

ANGLO-SAXONS – BRANZINGTUNE

The three centuries of the Roman period was a peaceful period for the people of the plateau. However, their lives depended entirely on events in Rome. Their rulers were always temporary ones, in spite of the army's policy of settling old soldiers among the British when their service was over. The moment came, about AD 420, when the legions finally went, recalled to defend the heart of the empire, leaving the British to govern themselves. Left alone, however, the people were no match for the next wave of invaders. These were bands of warrior-farmers, leaving their villages in north Germany in search of new land and settling throughout Britain. The ones who pushed north up the Street were Angles from Schleswig-Holstein. They had known all about Britain, its miles of woodland, ploughland and pasture, for many years. The Roman army had recruited their young men, valuing them for their love of fighting, and many of these Germanic soldiers had served in Britain. The departure of the legions, who usually had the better of pitched battles with the tribesmen, was the signal for migration.

The first of the Anglo-Saxon immigrants came into Derbyshire up the river Derwent soon after the Romans had gone, and settled in the valley. It was presumably later than this, perhaps during the following century, that the uplands to the south were occupied. The newcomers were organised in family or village groups. Two such groups settled and established themselves at Hopton and Carsington. A group led by a chieftain who may have been called Brand arrived some time in the fifth or sixth century on Carsington Pasture and, having surveyed the area and put the decision to a meeting of the whole group, in the usual way of these people, chose to settle in the valley to the west of the pasture. It was rocky and steep-sided, but it had over-riding attractions for Brand's group. They found springs, running off the limestone strata where they met impervious volcanic rock, into the lower ground of the valley sides, forming a stream which ran for a mile at the bottom of the valley.

They had the tools and the experience to clear a site big enough to build the very simple type of rectangular huts they had built in Germany. The present layout of the village suggests that they chose a site at the upper, northern, end of the valley, below the edge of the plateau, inconspicuous, sheltered and compact. The people seem to have made their settlement in a circle on the western slope of the valley, on ground which slopes west to east and north to south. The name of the area, Maddock Lake or, in the seventeenth century, Maggot Lake, suggests that the site enclosed a spring which ran sufficiently strongly for the people to construct a permanent water supply – in Old English a *maddock* was an earthworm or maggot and a lake was a small stream. At their backs, north, east and west, were rocky slopes and they set about clearing the ground in front of them, sloping to the south. They established pastures and meadows near the houses for the animals they had brought with them. Further down, in the centre of the valley and, where the land levelled out to the west, they set their ox-teams to work ploughing. Their ploughs were heavier and more effective than those used by earlier farmers. They travelled to and from their neighbours' settlements at Bradbourne and Carsington by the obvious routes to the south-west and south-east, while the Street dwindled to a track over Carsington Pasture.

Whether the British resisted Brand's and the other invading groups and were massacred, or whether they simply took them as new masters, the fact is that the only local village with a British name is Parwich. The settlements at Roystone, Rainster and Harborough are known only from their excavated remains, while the Anglian settlements were flourishing at the time of the Domesday survey and are still here. Brand's people may have found few in the valley to contest their arrival, and there was probably in any case little resistance to these natural fighters, arriving in force and determined to settle this new country. The valley springs may have been used by British groups, and broken pottery from Roman times has been found in the village, but the invaders were probably the first to settle in numbers in the valley. There has been no excavation in the village or in the fields to the south, but it seems likely that something more than a few shards of pottery would have surfaced over the centuries if there had been substantial settlement before the Anglians came.

The Anglians were pagans, worshipping the old German gods whose names we still pronounce every week – Woden's Day, Thor's Day, Freya's Day. They seem to have used the ancient burial places which they found near their new village. One of their leaders was buried during the 600s in the grave at Galleylow which was already 2,000 years old. This chieftain wore an elaborate necklace which consisted of a twisted spiral of gold wire joining thirteen gold pendants, eleven of them set with garnets. This ornament is similar to others found in Anglo-Saxon burials as far from Brassington as Kent, and demonstrates the trading and cultural links between the new communities throughout the country. The Galleylow noble fastened his cloak with an ivory pin, and his grave contained beads and pottery fragments. He was a warrior, buried with two of his iron-tipped spears, and his people were obviously expecting him to live on after death when they provided him with weapons and pots.

Brand's people clearly traded with their neighbours and with more distant Anglo-Saxon settlements. They were, however, forced to work out their own survival in a hard landscape and a harsh climate. The carpenters built rectangular wooden houses, with thatched roofs and a door in each of the long sides. These houses were a far cry from the villa, by then in ruins, which their neighbours had found at Carsington, but the people had built wooden houses in their old homelands, not to mention the boats which had brought them over the North Sea, and they were quite capable of building weatherproof houses in their new home. The houses had fireplaces for warmth and cooking, though there were to be no chimneys in ordinary houses for centuries – the smoke simply hung about the house until it escaped through the thatch. The group would include men who were skilful woodcarvers, and some of the houses would have decorated lintels and eaves. Furniture was as simple as chairs and tables could be. The people lived on oats, bread and vegetables, with occasional meat from wild animals, which included deer and boar, and from pigs and cattle, the latter slaughtered after their days as milk cows or draught animals were over. Grain, which included wheat and barley, was stored in large earthenware pots. The importance of barley is shown by the fact that the name for the building in which it was stored became the general name for a storage building – barn.

As they cleared the ground to the south of their village, and brought it under cultivation, the people distributed it among themselves in strips, taking care that good and poorer land was fairly shared. As the years passed and their numbers grew, the people enlarged the great fields in which the strips were grouped. As succeeding areas came under the plough they were shared out in the traditional way – every man had his share of strips, which the people called 'lands', and the work of tending the crops and harvesting them was organised on a village basis. This was a system which ensured something for every family in the village, but open fields and intermingled narrow strips made for quarrels and required elaborate rules to regulate the farming. Disputes over boundaries, rights of way over neighbours' strips and straying cattle, were settled at meetings of the whole village. To avoid violent quarrels endangering the village itself, the meetings were held near the Bradbourne border, where the road crossed a hill. The fields there are still called Spellow – 'the hill where speeches are made'– or, more often, Speller.

The present pattern of the fields near the village is plainly based on these original strips, and suggests that the villagers advanced the cultivated area down each side of the valley. The steepness of some of the fields on the eastern side indicates that the most unlikely land was tilled – every yard was needed to support the village. Halfway up the eastern slope they made their road to the south, probably running along the edges of the fields which were ploughed as far up the slope as was possible for the ox-teams to work. The strips ran down to the stream in the valley bottom, up the western slope and then further west, as the land levels out. The people brought about 500 acres into use as arable, meadow and pasture; enough, with the support of the surrounding wasteland, for game, timber and grazing to support a population of approaching 100 people. It is likely, though there is nothing to prove it, that the expansion of the fields to the south was rapid. The stream in the valley bottom is very shallow now, but there was a line of springs stretching the length of the village and used until about seventy years ago, suggesting that it had a much more powerful flow when the valley was first occupied. The dip in the road at the south of the village was often flooded up until the nineteenth century and the rules made by the manor-court in the sixteenth and seventeenth centuries stress the importance of keeping ditches clear. The land was well watered.

There are no records for Brassington earlier than the Domesday Book of 1086, but the period between the first Anglo-Saxon settlement in Britain and the Norman invasion saw a gradual shifting and coming together of communities through local and regional struggles. The people of Brassington no doubt cursed every interference with their lives caused by the struggles between first local leaders and later regional chiefs which culminated in the formation of the kingdom of Mercia near the end of the sixth century and, two centuries later, of the kingdom of England. Of equal importance to these political upheavals in its effect on the villagers' lives was the coming of Christianity. The second and great king of Mercia, Penda, who was killed in battle in 655 when he was over eighty years old, was a pagan. His son, however, married a Christian princess from Northumbria, and with her came missionaries and the conversion of the kingdom. Remote villages like Brassington were probably slow to change their ways, but the nobleman buried at Galleylow was one of the last pagans. His successors of the late seventh century and their people were Christians.

For the villagers, every political development brought a worsening of their lives. By the time of the Norman Conquest, they had long been working not for themselves but for a lord. While they had always been liable to pay a tribute of grain, cow hides, wool or anything else required of them by the current overlord, they were eventually forced into permanent servitude. The ground so laboriously won and cultivated for generations became the personal property of the lord. The villagers, or 'villeins', held their land at the lord's whim and worked the lord's land as well as their own. The origins of this 'feudal' system are unclear and the earliest records of its working in Brassington date from the fourteenth century, but the Domesday Book entry for the village is sufficient evidence that it was firmly in place by the time the Normans came. The land settled by Brand and his followers had become the 'manor' of Brassington.

THE MANOR

Brand's people, or their descendants, controlled an area of about 4,000 acres, stretching from the edge of Carsington Pasture to beyond Rainster Rocks, and from a mile below their settlement to three miles to the north of it. Where the valley ends the land rises gently to the south and the southern edge of the modern civil parish was probably the boundary of the English 'vill' or township from very early in the Anglian occupation. The boundaries of the vill were probably defined by the boundaries of the land-holding of the English lord who held the estate. There is a tenth-century charter of King Edgar granting neighbouring Ballidon to an English lord.[19] The boundaries of Ballidon then, defined in the charter, are the same as today's civil parish, including the boundary with Brassington. The medieval manor of Brassington extended north as far as Ivonbrook Grange and to Pikehall in the north-west, and it is probable that the Normans observed the English manor boundaries when they took it after the Conquest of 1066, as they clearly did in Ballidon. The manor included Aldwark, as did the township, the administrative unit – Aldwark was still being included with Brassington in tax assessments in the eighteenth century. The wasteland was common to the people of Brassington, and to no others – when the people of Elton and other villages were granted grazing on the moors they paid rent for the privilege. Boundaries were important and caused disputes – a manor-court on 9 November 1433, for instance, adjudicated on a boundary dispute between the '*tenantes Villat*' of Brassington and Carsington, which included possession of cornfields.

Without the documents to show how it happened, it can only be guessed that inequality within the village developed from the ordinary fortunes of life. Farming is precarious when there is little reserve to cover the inevitable bad weather and bad harvests. Many families must have been unable to continue to make a living from the land and thus forced to surrender it to more fortunate neighbours who could maintain them at the price of their labour. Outside the village, the emergence of larger political units meant greater demands for taxation, greater bureaucracy and greater call on the men for military service. The gap between rulers and peasantry grew and the villeins became tied to the land which they now held only in return for service to the nobleman who owned it. They were obliged to plough the lord's personal farmland, sow it and tend and harvest the crops. They mowed the lord's meadows, sheared his sheep, and supplied him with fencing, corn and ale. After all this they paid rent for their own land. When a villein died, his heirs paid the lord an 'entry fine' for his permission to hold the family land and livestock. These burdens varied from manor to manor. The village meeting became the meeting of the manor-court. Among the court's tasks was to provide the official who supervised all the activities of the villeins, the reeve. Harrowing, manuring, sheep-shearing, fencing, ditching, reaping, harvesting, ploughing, building winter shelters for cattle, timber cutting, garden and orchard planting and tending were all the responsibility of the reeve. He also saw that the village was supplied with ploughs, harrows, spades, scythes, sickles, shovels, rakes, and with pots and pans, barrels, bins, candlesticks, buckets, churns and every other tool and utensil needed. The manor was an economic and social unit, a tight and self-sufficient community.

VIKINGS

Two centuries before the Norman invasion the villagers suffered the same sort of calamity their ancestors had inflicted on the British, though this one had a much smaller effect. Scandinavian invaders – Vikings – had so harassed the newly established English kingdom that the king, Alfred the Great, had first been defeated by them and had subsequently, after regathering his forces and defeating them in their turn, in 878, made a division of the country with them. Brassington was in the Viking part of the country, the Danelaw, and the old Roman town of Deventio became the Danish centre Deoraby. Wirksworth, four miles from Brassington, became the centre of the Danelaw administrative unit called a *wapentake*. The influence of the Scandinavians remained strong in England and there was for a short time a Danish dynasty on the English throne. This was Canute, king from 1016 to 1035, and his sons Harold and Hardicanute.

In name at least, the Danish administration left its mark – Wirksworth wapentake kept its name for centuries, long after the English name 'hundred' had been substituted for it in similar units in Derbyshire. However, the English in Brassington seem to have been undisturbed by Danish settlers, if the scarcity of Danish place-names is an indication – the only indisputably Danish name in a list of the wastelands around Brassington in the seventeenth century is Jordan Slacke. [20] A slacke was the Danish name for a hill slope or hollow, and there are many with the name in Derbyshire. That there were no Danish settlers in Brassington is also suggested by the fact that the population described in the Domesday Book in 1086 included no 'sokemen'. These were free land-holders: *soke* was a Danish form of land-holding in which the tenant owed virtually nothing to his lord but loyalty, and there were substantial numbers of sokemen in all the main areas of Danish settlement. However, though the villagers carried on their lives undisturbed by Danish intruders, they had exchanged an English overlord for a Danish one, and when England reverted to native English kings, after the death of King Canute's son Hardicanute in 1042, the Danish overlord remained. By 1066 the lord was a nobleman with the Danish name of Siward.

REFERENCES

1 NA DL43/1/19.
2 Farey, vol. 3, 1817.
3 Hardy, 1762 (DLSL, No.12961, p. 22, MS note).
4, 6, 7, 8, 9 & 10 Marsden, 1977.
5, 11 & 12 Hart, 1981.
13 & 15 Branigan, 1986.
14 Ling, 1981.
16 Ling, 1990.
17 Hodges, 1981.
18 Dool, 1976.
19 Brooks, 1984.
20 Ince, 1871-2.

CHAPTER 2

MEDIEVAL BRASSINGTON

DOMESDAY BOOK

When William the Norman conquered the country he found a state with a highly organised local government system and a subject people tied to the land they farmed. He proceeded to install his own men to run the system, and to tighten it. The system ensured maximum profit for the nobility and a reliable supply of men and arms for William's armies. From 1066 the villagers at Brassington found that, while their village retained most of its old customs and ways of self-government, it was now one of the many manors owned by the Norman Henry de Ferrers.

They cannot have noticed much difference at first from the rule of Siward and his predecessors, as their new lord made no alteration in the running of the manor. They continued to take their orders from the reeve and his fellow officials, still working ceaselessly in their fields, still paying their dues in produce and service. In 1085, however, they received a clear indication that a new kind of ruler was in charge of the kingdom. A party of commissioners arrived in the village, one of dozens sent by the king to every county and every hundred in his new kingdom to gather information. The commissioners asked the same questions in every village and relied on the same group of people for their answers. These were the reeve, the village priest and six villagers. The answers they gave were checked for their accuracy by a second, different group of royal commissioners later in the year, and were incorporated in the Domesday Book. What the commissioners asked of the reeve, the priest and the six villagers was the name of the village, who had held it before 1066, who holds it now, what is its area, how many ploughs have the lord and the villagers, how many villeins, bordars, slaves and freemen there are, how much woodland, meadow and pasture is there and what was the value of the place before 1066 and what is it now. Written up in Latin in the king's great book the answers are translated as:

> In Branzincton Siward had four carucates of land assessed to the geld. There is land for four ploughs. There are now three ploughs in demesne and sixteen villeins and two bordars have six ploughs and thirty acres of meadow. There is underwood three furlongs in length and one in breadth. In the time of King Edward it was worth six pounds. Now it is worth three pounds.

The Domesday Book was partly a taxation document, intended to identify taxable land and buildings, partly a list of the king's vassal land-holders and their holdings, arranged under the names of the land-holders, and partly the best description of his kingdom that William could get, after he had spent most of his reign fighting a series of wars to consolidate his hold on it. Since the commissioners' scribes copied down the villagers' oral evidence, we can see that the village's name still, in 1085, echoed the name of its founder – Brand's people's town had not yet become Brassington. The four 'carucates' meant that the village included four taxable units of land, each of which was the amount of land which could be cultivated in a year using one plough. Clearly this amount depended on what sort of ground was being cultivated, and in fact varied widely throughout the country, from as little as 60 to as much as 180 acres. In later centuries the measurement at Brassington was 120 acres. The pre-Norman extent of the village's farmland may therefore have been 480 acres. How much of this was the lord's home farm – 'demesne' – is not stated. Perhaps its three ploughs compared with the six of the villagers means that it was a third of the whole, or 160 acres. The six ploughs, a term which included by implication the ox-teams which drew them, owned by the sixteen villeins and two bordars, would represent 320 acres. The population of the village, then, consisted of eighteen men and their families who between them held about 320 acres of ploughland and in return for this cultivated a further 160 acres of the lord's demesne.

Most of the 320 acres would be held by the villeins since bordars, being even lower on the economic scale than villeins, had only the plots on which their cottages stood. The 30 acres of meadowland for hay crops and grazing implies cattle and perhaps sheep, as well as oxen, though details of livestock, included in the commissioners' questions, were omitted when the replies were written into the Domesday Book. The 'underwood', an area also of 30 acres, was valued not only as a source of timber for building, for ploughs and harrows and a hundred other uses, but also as pasture for pigs and as a home for game. It was an essential part of a small village's territory. The halving of the value of the manor, its annual taxation value, was normal for the period – after the wars of the Conquest, following 1066, most English manors suffered a drop in value, and there were similar fluctuations in succeeding centuries.

The number of ploughs, the amount of arable land, meadow and woodland, and the number of men, mean that the village was thriving. Notably missing from the Domesday entry is any mention of a church or of lead mines. The existing church was built in the following century, but it is most unlikely that there was no church there by 1085. It was presumably exempt from tax. Similarly, it seems unlikely that the eleventh-century men of Brassington did no mining. '*Plumbariae*', or lead mines, were included in the entry for Wirksworth, and it may be that though there were mines at Brassington they were considered to have insufficient output to be taxed.

The Domesday Book mentions no freeholders in Brassington. It seems that everybody living there in 1085 was unfree. A villein was part of the lord's property and there is a matter-of-fact example of what this meant in practice in a document from about 1300, preserved in the Gell manuscripts in the Derbyshire Record Office.[1] This describes how the villein Ralph fitzRalph, together with all his goods and chattells, was given by his master, John de Aldewerke of Ible, to Dale Abbey. The land which a villein held in the village brought many other obligations as well as the obligation to work the lord's own land. He paid fines when his children were married and if his son was sent for schooling, since these were privileges granted by the lord. He paid a fine when he was 'admitted' to his dead father's holding, since it had reverted to the lord on his father's death, and a 'heriot', which was in Saxon times the dead tenant's military gear, supplied originally by the lord. By the time of the Conquest this had become the tenant's best beast and was eventually to become a money payment. A villein paid fines when he took wood from the wasteland, he paid a portion of the price of any beasts he sold and he was forced to take eggs to the manor house if he kept hens. He was constantly reminded of his servile status, and he could not leave the manor in which he was born. This was the system, sharpened and strengthened by the Normans, and some variation of it was in operation at Brassington.

The Domesday Book is the oldest surviving document mentioning Brassington, and nothing like so informative a statement about the village was to be made for more than 200 years. Every document during these 200 years had the same purpose – recording land ownership – but the Domesday Book, in listing amounts of land, numbers of ploughs and numbers of people, gave a picture of the village, meagre as it was, that was better than anything produced before the taxation documents, accounts and manor-court rolls of later centuries. There are charters and fines from the twelfth and thirteenth centuries which record grants of land or grazing right in Brassington. They enable us to see who held the land at different levels, but little else. A manor was a commodity. Its value to its owner lay in what he could get out of it in money or service. In recording its ownership, a charter tells us the name of the lord, sometimes his lady and the name of the sub-tenant he is granting part of his manor. It often names the area, which throws light on the extent of the manor, and where grazing right is involved, gives numbers of cattle or sheep. There is of course nothing about the people of the village.

LORDS OF THE MANOR

For 200 years, the tenant-in-chief, holding the manor of Brassington, and many others, directly from the king, was the head of the Ferrers family. The first was Henry de Ferrers, a Norman knight whose service to William at Hastings earned him the lordship of much of southern and central Derbyshire, and of Staffordshire. The family gained further promotion in the following century when Robert de Ferrers was created Earl of Derby by King Stephen after the Battle of the Standard in 1138. Half of the manor of Brassington was granted later in the century to a female member of the family and descended from her to the Furnivall family, who held it until the fourteenth century.[2] In that century Joan, the heiress of William, Lord Furnivall, married Thomas Neville, and in 1409 the heiress Maude Neville married Henry Talbot, who was created Earl of Shrewsbury in 1442 for his service in the French Wars. The Ferrers' connection with Brassington ended in 1266 when Robert, the 6th Earl of Derby, chose the wrong side in the Barons' Revolt against Henry III and, after beating a royal army at Chester and being imprisoned briefly in the Tower of London, was defeated at the Battle of Chesterfield.[3] Surprisingly, he escaped the fate usually suffered by those who had lost a battle at the time and was spared execution. However, all his possessions were forfeited and his Derbyshire and Staffordshire manors were among the territory granted to King Henry's son Edmund, who was created Earl of Lancaster. The earldom became the duchy of Lancaster when Edmund's son was created duke in 1351 and the duchy passed to John of Gaunt in 1362 after his marriage to Blanche, the sole heiress. It became Crown property when John of Gaunt's son led a rebellion against Richard II, won the kingdom and became Henry IV in 1399. For centuries this half of the original Ferrers manor was known variously as the King's or duchy-manor. The splitting of the original manor, and later developments in both halves, led to Brassington's future character as an 'open' village, that is one in which land ownership was widely spread, unlike neighbouring Tissington, for instance, owned by the Fitzherbert family.

It is most unlikely that the villeins and bordars of Brassington ever saw Henry de Ferrers or any of his successors. They may have been better acquainted with his sub-tenants, some of whom are known from surviving charters, but their immediate taskmasters were the ones they knew best — the lord's bailiff, who collected their rents, the steward who presided over the meetings of the manor-court which they had to attend, and the reeve who saw to it that they ploughed the lord's fields and harvested his hay and corn.

ABBOTS' LANDS AND VILLAGERS' RIGHTS

Among the sub-tenants whose presence was felt in the village were several of the religious houses who were rich, powerful and influential in pre-Reformation England. The Ferrers family made three grants of land to the Church. Great lords felt the need for regular prayers. They lived in highly dangerous relationships with each other and with their overlord, the king, and exploited the villagers living in their territories. They were hard men, but they were prepared to pay for religious insurance. Some of them were, no doubt, genuinely pious. For whatever motive, the first earl set up a convent in Derby during the reign of King Stephen, about the middle of the twelfth century.[4] This was an Augustinian house, and Earl Robert installed an abbot and monks and financed it from the revenues from some of his possessions, among them 6 acres in Aldwark, which was described in 1275, when Edward I's *Hundred Rolls* recorded land-holdings, as being 'of the fee of Bracinton'.[5] The convent was moved during the reign of Henry II to Darley, and the grant confirmed successively by Earl Robert's son and grandson, both called William.[6] The confirmation published by a charter of Hugh de Ferrers in the next generation includes the right of the monks to pasture 1,500 cattle 'through the whole moor of Bracinton.'[7] This charter reveals that the monks had been taking more than they were entitled to by the Ferrers' grants and attempts to put a limit on their use of Brassington and Aldwark, 'so that they shall not be able to put more cattle there in the name of or on account of their property.' Brassington Moor, stretching as far as Pikehall, was the precious wasteland of the manor, precious to the villagers for pasture and timber, and the rights which the lord had granted to outsiders in order to ease his conscience were rights which diminished the villagers' own.

The Ferrers family made two other grants to monasteries. Hugh de Ferrers granted pasture for 500 sheep in Brassington to the abbot and monks of Buildwas in Shropshire, a grant confirmed by the Earl of Derby in 1251.[8] A third religious house, Dale Abbey, held land at Griffe.[9] This consisted of 40 acres 'in the territory of Brassington lying between Hootlow (Atlow) and the King's highway leading to the mill of the monks at Ivenbrook', and pasture for six oxen, two cows and either 100 ewes and lambs or 200 sheep in Brassington. The 'monks at Ivenbrook' were the monks of Buildwas Abbey. The amount of the monks' holding is given in the *Hundred Roll* of 1275 as one carucate, or probably 120 acres, at Aldwark, held by Darley Abbey, and 80 acres in Brassington held by Dale Abbey. Griffe was valuable to Dale Abbey, and not only for its grazing: it was part of the Brassington ore-field and the monks were able to claim 'lot', the tax due to the owners of the mineral rights, from the miners. In practice they leased their rights to local gentry. There is a lease of the lot at Griffe to Richard Blackwall[10] in the Gell papers in the Derbyshire Record Office and another, of the lead duties and land there, to Ralph Gell in 1537.[11] At the dissolution of Dale Abbey soon after this, Gell paid the king £107 13s 4d for the abbey's land at Griffe[12], and the mines there remained as a private liberty, owned by the Gells.

These transactions demonstrate the vulnerability of the villagers to actions over which the system allowed them no control. For his own reasons, the lord allowed outsiders to encroach on land vital to the village economy and the outsiders clearly pushed their concession as far as they could. The charter of Hugh de Ferrers, limiting the monks' rights, must have been prompted by complaints from the villagers. Similar complaints about overstocking on the moors and encroachment on village land were to be made in succeeding centuries. Late in the fifteenth century the duchy tenants petitioned the Chancellor of the duchy against Ralph Eyre, a tenant of Dale Abbey.[13] Eyre held a piece of ground called Ernst, on the north-west of the village. The tenants complained that he had not only been pasturing his own cattle on duchy-manor territory but letting it to others, who were currently pasturing 500 sheep there. Eyre had also confiscated the tenants' cattle, 'to the utter undoing of the said tenants'. The duchy tenants begged that they and the tenants of the Shrewsbury Manor 'may have remedy in this behalve'. A little later, in 1517, the duchy appointed a commission to investigate complaints that Edene Bafford and William Old had been overstocking Brassington Common.[14] Testimony was given by villagers from Elton and Brassington that 'oute of tyme of mynde' the people of Elton, Carsington, Ible and Winster had had grazing right on Brassington Moor, for which they paid 13s 6d annually, plus one hen from every Elton householder with rights on the moor. In fact, a manor-court as early as 1361 had recorded a twenty-year grant of grazing right to Ible. The Brassington men giving evidence in 1517 were the leading men of the village – Richard Wylcokson, John Tyssyngton, William Bukstones, Thomas Tyssyngton, Robert Lande [Lane] the elder, Thomas Lande, Henry Gretton, Thomas Elen, John Lane, John Scattergood, Thomas Wryght, Thomas Gretton and Richard Sowter. A manor-court held on 25 April 1526 charged Elton and Aldwark 6s 8d for 'rasyng [animals] and pullyng [taking timber or game]' on Brassington Moor. John Fitzherbert and Adam Barsforth, presumably the same 'edene Bafford' of 1517, were charged 19s 6d for their use of the moor by the same court.

Brassington Moor was clearly valuable pasture land, though the grants all included arable as well as pasture rights. The Church had further privileges in the village in the shape of tithes – tenths of produce.[15] Tithes belonged to a fourth religious house, the Priory of Dunstable. Brassington church, built in the twelfth century, was a chapel of the church at Bradbourne, and Bradbourne church had been given to Dunstable Priory by the De Ferrers' tenant at Bradbourne, Geoffrey de Cauceis, in 1205. This meant that Dunstable was entitled to the great tithes, or those arising from the hay and corn, at Bradbourne and in the villages in which its chapels were established – Atlow, Ballidon, Tissington and Brassington. The value of the Brassington tithes is shown by the fact that when, in 1278, the priory chose to farm them out to a certain William de Hamiltune, he made a down payment of 200 marks, or £133 6s 8d. A quarrel ten years later between Dunstable Priory and Dale Abbey, where the abbey was alleged to have avoided paying tithes on its land, was settled by the abbey acknowledging the rights of the priory and the priory granting them to the abbey on payment of half a mark (6s 8d) per year, so long as the land lay under the direct cultivation of the abbey, and that no 'alien' animals were allowed pasture there. There were similar quarrels and settlements between Dunstable and Tutbury Priory, which had been granted tithes at Brassington in the eleventh century.

THE VILLAGE CHURCH

Brassington church was a fairly recent building at the time it and its mother church were given to Dunstable Priory. It was one of a number in Bradbourne and neighbouring villages, all with similar low, square towers, probably designed by the same master mason. Mugginton, Hognaston, Kirk Ireton, Bradbourne, Tissington, Brassington and Thorpe have churches with a strong family likeness.[16] Brassington's steeply sloping hillside posed problems which the builder solved by buttressing the tower on its south side and building an unusually narrow south isle.

Here he built three one-stepped arches, supported by heavy circular pillars with square, deeply carved capitals. As he had done at Kirk Ireton, the builder made the arches asymmetrical. The door was in the tower, which had, and still has, two-light bell openings. When he built the tower the master mason incorporated in the wall of the belfry a stone from Brassington's past, perhaps from a Saxon church, perhaps from the pre-Christian village. This stone has the crudely carved figure of a naked man with his hand on his heart. The builder made a round arch from the tower to the nave and, at the eastern end, a feature unique to this church, a continuation of the south aisle into the chancel. This extra arch has an octagonal pillar. The villagers started their lives with baptism in a large font built into the south wall.

The church was modified quite early in its life. The present entrance via the porch on the south may have replaced the tower door during the 1200s. The clerestory windows giving daylight to the nave are Perpendicular in style and may in fact have been installed during that period – the fourteenth to sixteenth centuries. The high pointed

St James' church, Brassington, from Church Street.

central arch to the chancel dates from the 1300s. These judgements on the age of the tower, nave, south aisle and porch are those of experts in church architecture, including Dr Cox[17] and Nicholas Pevsner's research team.[18] If they are correct, the villagers have had a church for 800 years, very little changed internally until the addition of a north aisle and an increase in the size of the chancel in the nineteenth century. The fact that the church was owned by Dunstable Priory, and was not part of either manor, has meant that there are no records surviving of early repairs and alterations. Its history is to be deduced from the building as it is now.

The church, and its lost predecessor, were central to the villagers' lives. They were baptised, buried and, after the Reformation, married there. Their holidays and festivals were Church celebrations and the solemn services, though conducted in incomprehensible Latin, supplied a spiritual side to lives which were otherwise spent in long hours of hard labour. It is likely that throughout the medieval period there were no seats for the congregation, and the villagers celebrated the rites standing in a huddle in the nave. The church was the most substantial building in the village, visible evidence of the power of religion. Here were the only pictures most of the villeins would ever see, and here their only instruction as they listened to the parson's sermon or a fiery address from a visiting friar.

VILLAGERS

And while the king's tenant-in-chief was using his manors, including Brassington, to fill his treasury, exact his service, reward his friends and save his soul, playing the high politics which eventually lost him his inheritance, what of the men and women who did the work on which it all depended? The sixteen villeins and two bordars of the Domesday Book are anonymous and their descendants remained unnamed for more than two centuries. Their assembly had by the eleventh century become the lord's court. They were obliged to attend for the roll call and acknowledge their readiness to do 'suite and service' to the lord. This phrase persisted in the ritual of the manor long after it had lost its meaning, but this meaning was real enough in the days of the Ferrers' courts and long after their time. The presence of the villagers and their services and payments were essential to the running of the manor. The court recorded their land-holdings and the service and, later, the payments which they owed for them. It recorded their misdemeanours and the fines they were charged for them.

In Brassington these records have survived from 1300, by which time many of the villagers had begun to be known by surnames as well as the Christian names they had always had – until the fourteenth century, the Johns, Williams, and the Margarets and Ediths of the village had no family names and were simply John son of William, William of Ballidon, Roger the shepherd or, like a sub-tenant in one of the charters, Hamo the clerk. It was during the fourteenth and fifteenth centuries that family names became stabilised. The earliest surviving court roll, one for 1300, has a mixture of modern surnames – William Osborne, Robert Adams, William Foy – and the old names based on fathers' names or on place, such as John son of John, John del Hull, Roger de Brabazon. Slightly later rolls have Henry à la Dale, William de Bryth de Braston (Brassington) and two names which tell us something about the village. Henry Bercar, or Henry the shepherd, reminds us of the earl's sheepwalk, though he may have been employed by Dale Abbey or Darley Abbey to tend the flocks which they pastured on Brassington Moor. Henry Molend – Henry the miller – presumably carried on his trade at the mill mentioned in a number of manorial accounts.[19] The mill was probably a water mill, powered by the stream at the bottom of the village. The court records also identify fields whose names persisted for centuries – Ragstaffe, transferred by William Elyn to his son William in 1379, or Caldwell Syche, occupied by Ralph Mylnes in 1433, for instance. From 1300 we can see something of the lives of some of the people of the village.

One of the earliest rolls has an example of the keen interest the lord had in the most private of the villagers' activities – his livelihood depended on them and he superintended their lives at every point. When Edith, daughter of Roger Eleyne, fell in love with Julian, son of William Fisher, in 1301, they had to wait until Roger got the lord's permission before they could be married. The villagers found release from what must have been a constant sense of servitude and from their toil, in at least one time-honoured way – drinking. Even here the lord made money out of them. The Assize of Ale, passed by Parliament in 1267, had prescribed the strength of English ale, and a manorial official checked it. In almost every court roll, the village brewers are fined 3d for 'breaking the Assize of Ale'. There cannot have been such consistently poor ale and the 3d sounds suspiciously like an unofficial tax. Alehouses were often kept by women, and there were two in Brassington in the early fourteenth century. One of the two was Agnes de Ballidon, a Brassington woman despite her name. Her husband's ancestors no doubt came from neighbouring Ballidon, and he was probably the William de Ballidon mentioned in several court rolls. By the end of the century the name had become plain Ballidon and the family was still in Brassington three centuries later. Agnes's contemporary was Margaret Cole. There were five or six alehouses at all times, their owners regularly fined. Near the end of the century three of them were women – Cealia de Bradburne, Joan Filyote and Alice de Maston.

THE VILLAGE ECONOMY

Also surviving from the early fourteenth century are accounts kept by the Honour of Tutbury, the name given to the Earl of Lancaster's Derbyshire and Staffordshire estates. These throw light on the village economy by measuring the fluctuations in the amounts of land and property for which the villagers paid rents – during prosperous times the population rose and more land was taken into cultivation. An account for 1313 shows

Ridge and furrow patterns revealed by melting snow in the fields to the north-west, between Lot's Lane and Rainster Rocks. There are lead-mine shafts and hillocks in the foreground.

the position at the end of a long period of expansion.[20] There had been a rise in the amount of farmland in the village during the 1200s, farmland which most probably included land beyond the Bradbourne Road, won from the wastes. On the north side of the Bradbourne Road and among the rough ground between Lot's Lane and Rainster Rocks are the unmistakeable ridges and furrows left by the medieval ploughmen.

To the south of Rainster Rocks it is apparent that the villagers ploughed every piece of reasonably level ground – the ridges and furrows reach up to the rocks and the slopes below Lot's Lane. This use of every available square yard demonstrates the pressure on space caused by a rising population. The extension of cultivation into the wastes was known as 'assarting', and the 1313 accounts show that 60 acres were assarted in Brassington at that time.

A court roll of 1301 has a list of nine land-holders, which means that, assuming a similar number in the other manor, the land was divided among the same number of villagers as it had been in 1086. No one on this list has a great holding. The largest is Roger de Brabazon's 75 acres near 'Le Howe'; William Osborne and members of the De Howe family have a few acres each and the rest either one or two bovats, a bovat or oxgang being about 15 acres in Brassington. Three of the men have a 'messuage', or farmhouse, as well as their land-holding, which in these three cases is two bovats, putting the three men among the larger land-holders in the village. Another tenant, in addition to two bovats, has four acres of demesne land, indicating that at that early date the lord's own farmland was being let out to the villeins instead of being worked by them in return for their own holdings elsewhere among the 'lands'. The accounts confirm that by 1313 the whole of the demesne was being leased – it amounted to 33.5 acres of pasture and about 60 acres of arable. The total acreage in the court roll list is about 227, so that with the 93 acres of demesne, the duchy-manor lands totalled 370. If the other manor's land came to the same amount, the total acreage of arable and pasture in Brassington by 1300 was about 740, an increase of 45 per cent on the amount being farmed in Domesday times.

The early court rolls have no suite rolls, or lists of the manor tenants, but the number of names mentioned makes it clear that there were many more than the eighteen men of Domesday and their families in the village. The 1301 roll which lists the nine land-holders names two other men – one of whom, Roger Eleyne, was of a family which remained in the village for another 200 years. One of the Eleynes, named in a land transaction in 1363, was described as 'chaplain', presumably meaning that he was priest in charge of the Brassington chapel. As pointers to the size of the village population, rolls of 1308, 1327 and 1334 have another thirty-three names in total. Any estimate based on such shaky statistical foundations may be off the mark, but it does seem from these rolls that the village may by the fourteenth century have had 200 inhabitants. The fortunes of the villagers fluctuated, however, and the period of rising population and extension of farmland was interrupted in about 1320. Accounts for 1321[21] show that the villag-ers had abandoned 54 acres of assart land and 13.5 of demesne. It was probably at this time that the land at Rainster was abandoned, never again to be ploughed. The reason given by the earl's servants for this decline at Brassington was that the land was poor, and other reasons given for decline in the Honour of Tutbury generally were the pov-erty of the tenants, which prevented them from paying their rents and caused them to give up their holdings, and a shortage of livestock caused by disease ('murrain'). There had been a period of bad weather and poor harvests which had caused the so-called Great European Famine (1315-17). This must have affected both population and stock levels. The poverty of the soil was aggravated by over-cultivation. This would stimulate assarting, and the newly-cultivated ground would itself become exhausted after the first fertile years.

The slow pace of social change was suddenly accelerated in the fourteenth century by the arrival in England of the epidemic of bubonic plague called the Black Death. It reached this country in 1348, spread by infected fleas carried by rats in the ships trading across the English Channel. By 1380 the death rate was so high that the national popula-tion is estimated to have fallen by between a third and a half. At Brassington there are no court rolls between 1347 and 1361 – perhaps an indication that the plague had killed so many of the villagers that the normal routines of the manor could not be carried on. Later rolls do not mention the plague, but its effect can be seen in a Tutbury account for 1361-2, which records that cottages, assarts, arable and meadowland worth £3 10s, or almost a quarter of the rent roll, were unoccupied. Only one acre of the abandoned land was taken back into cultivation between 1361 and 1377.[22]

The effects of the shortage of labour caused by plague deaths were momentous for the surviving villagers. The lord's urgent need for tenants and labourers to work the manor forced a loosening of the system. Villeins were set free, their wages raised, and tenancies made as attractive to the villagers as the lord could afford. In Brassington there is no reason to doubt that these effects were felt. A rental of 1410[23] suggests that whole families may have died. It lists twelve men, eleven of whom have family names which were unknown in the village before the middle of the fourteenth century. There is only one name from the 1301 list. Seven of the 1410 tenants are from families who remained in Brassington for several centuries – John Gratton, Ralph Mylward, Richard Wysall, John Taylor, John Wheston, Henry Hicheson and William Wheston. The Whestons may have been the ancestors of the Westernes, who were the dominant family in Brassington by the seventeenth century, and it seems likely that these new families were benefiting from the urgent need on the manor for labourers and tenants which the Black Death had created.

Accounts for the fifteenth century show fluctuations in the amount of rent collected by the Tutbury officials. These fluctuations may in part have been due to changes in rent levels, but they must also indicate rises and falls in the amount of land cultivated. The century was a period of instability, including a long civil war, the Wars of the Roses, and of economic stagnation throughout the country, in which the village seems to have shared. An indication of the manor's continuing need to encourage tenants to take on holdings is an account in 1446-7 of grants made to two Brassington tenants to take holdings of five and three bovats which had been unoccupied.[24] The grants were of £2 and £1 for necessary repairs, and the tenants agreed to maintain the holding themselves after the initial grants. These efforts by the lord, who since the accession of Henry IV in 1399 had been the king, were only partly successful, and there was less land in cultivation at the end of the fifteenth century than there had been at the beginning of the fourteenth. The account for 1460-1, for instance, recorded that 12.5 per cent of the rent charge could not be collected because of land lying out of use. The total rent charge for Brassington Manor fell during the plague years, from £14 16s 4d in 1361-2, to £11 2s in 1484-5. In the last years of the century there was a recovery, but the rents never rose to the 1361 figure. The other manor seems, from the very limited information which has survived from the fifteenth century, to have been much less profitable – Henry Talbot, the future Earl of Shrewsbury, paid a 'knight's fee' of £7 for it in 1431[25], and an account for 1446 gives the annual rent as £6 26s.

The manor-court's powers must have seemed wide enough to the villagers at a time when central authority was relatively undeveloped and the local lords wielded a power in their own areas rarely checked by the king's own representative, the county sheriff. Only the most serious offences were tried at the sheriff's court, and it is unlikely that there were many in a community of only 200 people. Most of the villagers' misbehaviour was dealt with by the manor-court. These included such skirmishes as the one for which Henry de Chrulye was fined 12d in 1334 – for drawing blood in a fight with Henry the Miller – and William, son of Emma de Duggus, the same for wounding Henry Farson of Ible. The court also laid down the rules for the running of the open-field farming system, and fined the villagers who infringed them. They were often infringed, as was to be expected where the villagers' land consisted of scattered strips, intermingled with their neighbours'. There was always the danger of part of a man's crop being harvested by his neighbour, of animals being turned prematurely out to pasture, before the grass had recovered from the winter, or onto unharvested crops. The problems and the rules to deal with them remained the same throughout the centuries before the strips were amalgamated into small fields, and the great open fields disappeared. Just before the first large-scale enclosure the court ruled that:

> Item that [no] person gather pease but their own nor gaine anie corne in the fields but at theyr Barne dores upon payne of every tyme 12d.
> Item that noe person carry or recarry thorowe ye medowes before Michelmasse upon payne of every defalte 12d.
> Item that no person doe putt into ye fallowe any lambes before Saincte Barnabyes day next in payne of every day 12d.

The collective nature of this farming is demonstrated by another 'payne' made at the same time: 'Item that noe persons shall tether anie Beastes [cattle] or capulls [horses] but upon his owne grounde… without consent of the whole Towne upon payne of every tyme 4d.'

The successful working of an open-field system depended upon co-operation among the villagers and these rules demonstrate that human nature needed a little help from the authorities if the system were not to break down. It is not possible to work out from the pattern of strips still easily to be seen at Brassington whether the village operated a two- or three-field system. One field is known to have been Town Field, and it may be that there were in fact only two: Town Field, near the village, and an outer one. Whether two or three, the fields were cultivated in rotation, with one being left fallow each year;

manuring was accomplished by folding sheep and pasturing cattle on the grass and weeds of the fallow, and on the cultivated field once the crops had been harvested. The timetable had to be observed – 'no person doe putt into ye fallowe anie lambes before Saincte Barnabyes day next.' The number of cattle and sheep which a villager was permitted to pasture on the commons and wasteland, his 'beastgates' and 'sheepgates', was related to the amount of land he held – a system known as the 'stint'– and later lists of manor rules put the number of cattle per oxgang at four and sheep at sixty. Breaking the stint was one of the misdemeanours punished by the manor-court on many occasions. John Wallwyne, for instance, was brought before the court on 27 April 1462 for putting more cattle on the 'mora de Brassington' than his holding entitled him to. Access to pasture was of the first importance to the villagers, and at Brassington the situation was complicated by the fact that for centuries the moors were shared with the people of neighbouring Elton, Winster, Carsington, Aldwark and Ible.

THE HOUSES – OUTSIDE AND INSIDE

The villagers' living conditions changed little during the centuries. They continued to build their single-roomed, mud- or wooden-walled huts, roofed with straw or reed thatch. The Tutbury accounts for 1441[27] describe the building of a bigger house than most, consisting of four bays, a bay being a standard building length of four yards. There were low masonry walls supporting a timber framework which was filled in with wattle and-daub panels – brushwood and branches cemented by clay. The house had rubble-floor and a thatched roof. It was built at the manor's expense and cost £13 9s 2d, most of which went on the wages of the men who built it – John Wyberly, William Frome and Robert Whitgreve, who, from their unfamiliar names, seem to have been from outside the village and may have been specialist house builders. Some items were bought in from outside the village, including nails – 'lathnayles' at 4s 2d and 'bordnayles' at 6s 8d. A thatched three-bay barn was built at the same time.

From 1535 there are wills and inventories, or lists of possessions, to give an idea of the villagers' houses. The earliest were left by men whose lives had begun in the fifteenth century and whose style of living was clearly more like the style of preceding centuries than of the years of progress which started in the reign of Queen Elizabeth. Ralph Knollis (Knowles) for instance, who died in 1536, seems to have had such a small and sparsely-furnished house that his neighbours John Charlton and Robert Briddon devoted only one line of his inventory to his household goods when they appraised them. Only the better-off left wills at this period and Knollis was a member of one of the yeoman families in the village. Briddon had appraised the goods left by another village notable, Thomas Charlton, a year earlier. Charlton had more than Knollis but the picture is still one of great simplicity. He had three pots, four pans, three sideboards, six bedcovers, six blankets, twelve sheets, three coffers, four ale tubs, one table, and one form. The table consisted of a board and two trestles to rest it on, and most notable is the absence of a bed. The Charltons must have slept on the floor or on the coffers (chests). The inventory names no rooms and it is likely that in fact Charlton's house was single-storeyed and one-roomed. His neighbour, William Bucstone (Buxton), whose possessions included a hundred sheep and four 'bulocks', (probably oxen), had even fewer household goods than Charlton. He had what seemed to have been a luxury at that time, a mattress; but no bed. These were the most prosperous men in the village, whose family names were often mentioned in the court rolls as the inheritors or buyers of land, and as officials. Bucstone, for instance, was reeve in 1527 and 1534. Their spartan houses must have been the best in the village, most of the Brassington houses being even smaller and more sparsely furnished. Few people in the village had the sheets, coverlets and chests listed in these inventories and their possessions were too few to be worth recording.

Maddock Lake, at the heart of the village and the probable site of the first settlement.

Church Street.

Town Street.

The year 1485, when Henry Tudor ousted Richard III at the Battle of Bosworth, is generally reckoned to mark the end of medieval England. It certainly saw the end of the long rivalry between warring groups of nobles and the start of the strong central rule established by the Tudors. Neither countries nor villages change abruptly and the change of king in London must have seemed less important to the villagers than the annual change of reeve, or the arrival of the lord's steward to conduct a meeting of the manor-court. The old ways persisted, but there was to be such a transformation in the people's houses, in their ways of making a living and in the way they were controlled by their masters, that it is worth looking at Brassington at the end of the fifteenth century, to see what was to disappear.

The only building remaining in Brassington from the village of 1500 is the church, which was narrower then as the north aisle was not to be added until the nineteenth century. The tower remained unaltered during repairs and extensions in the eighteenth and nineteenth centuries and the general appearance of the church was not changed. Apart from the church, the village, instead of the present cluster of limestone buildings, still consisted of the rough, wooden-framed, mud and wattle houses, single-storeyed and thatched with straw or reeds. The village was smaller then. Kingshill, called Kinchill or Kinshill then, as it is still sometimes pronounced in Brassington, was still partly 'waste or common' at least as late as 1653, when it was so described in the manor-court book. The houses built in the early seventeenth century on Town Street, Church Street and the Hillside are likely to have been part of an expansion along the roads to Derby and Bradbourne. The original settlement in a circle formed by Maddock Lake, Town Street and Miners Hill suggested by the map of the village would have the houses in 1500 lining each side of these streets, with their crofts and yards in the middle.

A lane called Badger Lane in the sixteenth and seventeenth centuries was probably the one now called Jasper Lane, running from the western end of Maddock Lake to the wastes at Kingshill. The wastes stretched to the north across Brassington Moor and east on Carsington Pasture, beyond the sloping strip of common pasture called Sides which ran on the eastern side of the highway. A second and much larger area of pasture lay on the west – the Over Pasture. To the south, between the two highways, were the villagers' meadow lands, and beyond, spreading to the south and south-west, were the lands – the cultivated open fields. The wastes were fenced off from the meadow, pasture and ploughland, with gates at Lydgate in the south-west, Longcliffegate in the north, Badger Lane Gate and gates called Doddersfieldgate, Lamberfieldgate, Longdalegate and, at the south end of the village, Green Gate. The pattern of the villagers' lives was still, in 1500, similar to what it had been for several centuries. None of them had been serfs – 'bondsmen' – for a century or more, but the descendants of the bondsmen had powerful economic restraints on their lives. There was little money to buy goods from outside the village and they grew their own food, made their own clothes and built their own homes. There was work, hard work, for most, but change had begun by 1500, and the flocks of sheep, needing few men to work them, had already deprived some of the security of arable farming. The next century was to see the start of the transformation of the open fields into the smaller ones which still form the pattern at Brassington. The communal farming methods implicit in the open-field system were to disappear with the rearrangement of the strips into the small hedged fields still to be seen. When a man's holding was in half a dozen separate strips intermingled with his neighbours' he and those neighbours needed each other. The coming of compact, individually-owned holdings made for a different society, and it came in little more than a century. Within the same period the settlement expanded and the present limestone village began to appear.

REFERENCES

1 DRO D258/53/4

2 *Dictionary of National Biography*

3, 13 & 14 Blanchard, 1971

4, 6 & 7 Kerry, 1894

5 & 25 Yeatman, 1886

8 Jeayes, 1906

9 MLSL BM Add MS 6674 f195

10 DRO D258/53/1r

11 DRO D258/53/3a

12 DRO D258/53/3c

15 Saltman, 1966

16 Currey, 1903

17 Cox, 1877

18 Pevsner, 1978

19 Blanchard, 1967

20, 21, 22 & 24 Birrell, 1962

23 MLSL BM Add MS 6666 f271

26 SCA S112

27 NA DL29/369/6180 (quoted in Blanchard, 1967)

CHAPTER 3

OUT OF THE MIDDLE AGES

TUDOR CHANGE AND VILLAGE CONTINUITY

The sixteenth century, the Tudor century, and in particular the long reign of Queen Elizabeth, saw an end to stagnation in English life. There were basic changes; a great increase in trade and a rapid development of capitalism, a religious revolution and, at the end of the century, a great rebuilding. Throughout the country, townsmen and villagers built new dwellings and yeomen, prospering with the increase in the wool trade, began to furnish their houses to a degree which we can recognise as in some measure comfortable. There was little comfort before the Elizabethan age. By 1563 Brassington had grown to a village of sixty-one households, or between 250 and 300 people[1], and while the yeomen (usually meaning freeholders), husbandmen (usually tenant farmers) and lead miners who had replaced the villeins and bordars, the Buxtons, Charltons, Allsops, Knowleses, Lanes and Jacksons, lived their lives in a framework of custom and tradition which must have seemed timeless, their village, small and remote as it was, was changing with the times.

Custom and tradition was the business of the manor-courts, but the sixteenth century in fact saw the beginning of the transfer of power away from the manor to the parish vestry. The process started in 1555 when a Highway Act transferred responsibility for the highways from the manor to the parish, and was accelerated by Queen Elizabeth's Poor Law of 1563, the first of many attempts by Elizabeth and her successors to cope with the increasing numbers of poor people. The parish was given the task of appointing collectors of alms for the poor and accounting each quarter to the vestry. This decay in the power of the manor-court was gradual and for a century or more it is apparent from the manor rolls that the courts shared with the parish the tasks of maintaining the highways and feeding the poor.

The two manors continued in the possession of the duchy of Lancaster and the Earl of Shrewsbury respectively for the whole of the sixteenth century, and the courts continued to meet regularly to lay down rules, try villagers for breaking them, and to transfer land and houses. The court rolls give a picture of a living and busy manorial system throughout the century. There are fines for animals in cornfields as well as pasture, indicating that the open-field system continued to cause quarrels among the villagers – keeping animals where they should be was a complicated business. The manor had an official called the 'hayward', whose duty was to check on fences and seize any animal found outside its rightful place and pen it in the 'pound', maintained by the 'pinnar'. The pound or pinfold was in the centre of the village, near Maddock Lake. According to Mary Walker, born about the turn of the nineteenth century, the last stones from the pinfold were removed only in the twentieth century. There was a charge for recovering strays from the pound and a fine for removing them without the pinnar's consent: 'Item that noe person do break Quenes mte [majesty] common pound in payne of every tyme Xs.' The importance attached to the offence is shown by the size of the fine, 10s, and it continued to be seen as serious into the nineteenth century. Jacob Johnson was imprisoned in Ashbourne Gaol for a month in 1822 for 'breaking the pound.'[2] The land-holders bought and sold their property, or made sure their sons inherited it, by observing the proper formalities. For example, on 20 February 1542, Wiliam Buxton surrendered to the court a messuage, two bovats of land, a curtilage (the outbuildings of his farmhouse), Coole Well, eight crofts, Caldwell Syche and Daggegrene, and the jury agreed that after his death the property should go to his son John. The system seems, from the court records, timeless and unchanging.

In one respect the duchy-manor did change very little: the rents audited at Tutbury remained at £13 8s 10d for the whole period covered by a surviving account book – 1505 to 1542.[3] They had risen fairly rapidly in the later years of the fifteenth century because of a rise in the number of cattle pastured in the village and consequent pressure on grazing.

However, these were the peak years for the duchy-manor rents and indeed they seem to have been set too high, since annual arrears rose from nil in 1505 to £20 in 1540. By the time of rental in 1620[4] the total rents had actually fallen to £7 14s 10d, a remarkable drop since the reign of Elizabeth had seen a rapid price inflation. The tenants had achieved this by using the ancient machinery of the manor to improve their conditions.

THE END OF FEUDALISM IN THE MANORS

What went on during the sixteenth century was revealed by the proceedings of a case brought before the court of the duchy of Lancaster by the Attorney General of the duchy in 1620.[5] He alleged that the land-holders of Brassington and neighbouring manors had exchanged their strips in order to bring each man's strips together and enclose them, and that these transactions had not been brought to the manor-court. The villagers had had land transfers approved by under-stewards and other inferior officers so that they would not 'be prevented in their said courses by some order to be therein taken by this Court.' These illegal transfers had resulted in a loss of fines paid to the manor when land was transferred. The duchy also accused the tenants of illegally enclosing parts of the waste and building houses on them. A further charge was that the villagers in Brassington had taken a piece of demesne land called Reeve Holme, pretending that it was theirs. This piece of land was held each year by the annually-elected reeve to recompense him for his duties, which included collecting the lord's rents.

The tenants, twelve of the leading men of the village, met as the manor-court jury and put together their answer.[6] In doing this, they documented the end of manorial service and proclaimed their right to manorial land. The change had been happening for many years, but the tenants' answer to the duchy was a clear statement that the old days had gone. What they wrote was a 'custumal' or statement of the customs of the manor. Custumals were usually compiled as a result of a quarrel between the lord and his tenants, and this one claimed to be 'a full measure of these points' alleged against them.

The custumal, at the head of a roll of parchment made of several pieces sewn together, has lost a part and some of the writing has become illegible over the centuries, but enough survives to make the tenants' case clear. The roll begins, 'The Jury do...' and these hard-headed yeomen put their case simply and clearly. They say that the custom of the estate is to have their land in 'fee simple', which meant that they could sell or bequeath it to anyone they wished and that they regarded it as their own. They have rights of pasture and 'turbary' (digging turf) on the common land and owe 'ne works nor boones nor other duties' to the lord. They stress the poverty of the land at Brassington, where there are no woods and very few hedgerows and where lead mining lowers the value of the land and poisons the cattle with 'uncurable' belland. They say that the principal profit in the land is from the lead mines and point out that every thirteenth 'dish' of ore goes to the king or his 'farmer', the entrepreneur who has leased the rights to the mining dues.

The value of the common is diminished by rights which the people of Carsington, Ible and Aldwark have there and the Brassington men assert that the profit from enclosing part of it would be less than the cost of doing it – 'the inclosinge thereof would be more charge then groth could be had by it.' They bring in the weather, as true Englishmen: 'the clymate is so could in winter.' They deny that anyone in Brassington has been confusing manorial land with freehold, thereby losing the lord his rent, and assert that every transfer of land has been properly carried out in the manor-court. Reeve Holme had never been demesne land and should continue to be used by the reeve.

Each entry in the rental which follows contains these words: 'upon wch premises there is neither millne woods nor quarries, neither is the same harriotable nor anie services boones works or duties dew or done for the same, other than Reeving service & suite of court as by auncient custome hath been used.' Harriotable meant attracting the payment of a heriot to the lord on inheritance; reeving service was carrying out the duties of the reeve or other manorial official, suite of court was the duty of attendance at meetings of the manor-court. Services in kind, the original basis on which the villeins had held their land, had clearly been lapsed long enough for the tenants of 1620 safely to exclude them from 'auncient custome'. The services had been replaced by the annual rent, and the tenants' title to their land was a copy of the entry in the court roll which recorded its inheritance or purchase. They were 'copyholders'.

The case was settled by compromise[7]: the tenants' rents were fixed and their copyholds and privileges confirmed on payment of thirty-five years 'Cheife and auncient rent' in two equal sums, one to be paid within three months of the court's decision and the second three months after its confirmation by Act of Parliament. In Brassington's case, the payments amounted to £139 9s 7d. The court also conceded rights which included pasturage for sixty sheep and four cattle per oxgang of each holding, and permission to get limestone, slate, gritstone, gravel, marl, sand, clay, peat, turf, heath, ferns, firs, gorse and timber from the wastes. The final confirmation of the duchy court's decision was delayed by conflicts between Parliament and Charles I, and by the Civil War and the Commonwealth government. It did not come until the reign of Charles II, by which time the tenants were all dead and the manor sold out of Crown ownership. The tenants of the privately-owned manor were the beneficiaries of their ancestors' arguments, put successfully in 1620. What the tenants of 1620 had done was to confirm a situation which had developed over many years. The local officials of the duchy-manor had allowed rents to remain static, duties to the manor to lapse and demesne land and waste to be taken over. The 1620 decision meant that, so long as transactions were carried out in the manor-court, the copyhold land of the duchy-manor could be transferred by sale or inheritance to whomever the tenants wished, while the rents remained unaltered for very long periods.

While the tenants of the duchy-manor were using the complicated bureaucracy of the duchy of Lancaster to subvert the system and maintain low rents for their copyholds, their neighbours, whose lord was the Earl of Shrewsbury, were experiencing quite different changes: by the 1580s, almost all the land in this manor was held by leasehold, and the earl was extracting the maximum profit from his property by raising rents at the end of every lease. While the duchy tenants were paying about 4d an acre, year after year, their neighbours, whose strips were intermingled with their own, were paying ten times as much and forty years later twenty times as much.

FARMING

For most of the sixteenth century, the farming at Brassington continued on the open-field system. In the 1580s, the Earl of Shrewsbury made a start on what was to prove a fairly rapid process of enclosing his land and, as the attorney general alleged in 1620, the duchy tenants enclosed their land by exchanging strips among themselves, but until then the old methods persisted. The villagers grew the crops they had always grown; the inventories of men and women dying between 1535 and 1600 name hay and corn among their possessions. Only one man, John Lane, in 1597, is identified as growing wheat, but later inventories name wheat, oats, barley and flax. A letter written in 1575 by the Earl of Shrewsbury's bailiff, William Dickenson, now preserved in the Lambeth Palace Library, refers to 13 acres sown with 'hardcorne' (wheat or rye) at Brassington.[8] Peas are mentioned in a manor-court case of 1586, and land transfers often refer to villagers' 'hempyards' – they grew hemp in their gardens and made their own rope.

Their ploughs were simple wooden tools, as were their harrows – the valuations of these implements in their inventories make it clear that these were nothing like the expensive machinery of later years. They were made by the villagers themselves, perhaps by carpenters, though making a plough or a harrow was probably a skill posessed by the average sixteenth-century farmer. There is evidence that these ploughs and harrows had by the end of the century begun to be improved by the addition of iron at their cutting ends and edges. Hugh Crycllowe's goods in 1582 included iron 'culters and shares' listed separately from his 'waynes and plowes' – the appraisers were listing iron and wooden goods separately. The wains, too, had by the end of the century begun to have the iron tyres on their wheels, which were soon to become universal. John Buxton's inventory of 1574 lists 'one bouden (iron bound) weyne' and his, too, lists 'one culter one sherre' separately from his plough, and the fact that his harrow is listed with them may mean that it had been improved by the addition of metal tips, forged by the blacksmith in the village. These are the only implements mentioned in the sixteenth-century inventories, and the appraisers may have regarded such items as forks and sickles as being not important enough for mention. They were obviously part of every farmer's stock and in fact the miner/farmer Roger Jackson owned 'two pickforkes and two sickles', together valued at 8s, when he died in 1614. Three of the inventories include the yokes used to harness oxen to plough and wain and one, Hugh Crycllowe's, has 'iron teames', the chains used with both oxen and horses. None mention horse harnesses in spite of the fact that seven of the men had horses. It is likely that for most of the century the heavy work of ploughing and carting was still carried out by ox-teams.

The villagers relied on the bigger farmers to provide the oxen, in the same way that modern farmers borrow or hire the more expensive machinery from each other or from contractors. Only six of those who left wills in the sixteenth century could afford to own oxen, which were given the highest value of the village animals by the appraisers. Thomas Charlton's six were valued at 10s each in 1535, and by the end of the century the valuation had risen to 55s. This figure was not reached by horses until the middle of the seventeenth century.

Most of the farming throughout the sixteenth century continued to be arable. There are too few cattle in the inventories for them to have been reared for the market. These 'kyne' and 'heyfers' were for the milk and meat drunk and eaten in the village. Most of the inventories list a few sheep, and there were four sizeable flocks. William Bucstone had 100 sheep in 1541, Hugh Crycllowe 217 in 1582, John Buxton 60 in 1574 and John Lane 74 in 1591. Even these four were also arable farmers since all had ox-teams. In addition to the draught animals, cattle and sheep, the villagers kept pigs. Only two of the inventories list them but the concern of the manor-court that pigs should not go unringed shows that there were enough in the village to damage grazing land if they were allowed to grub freely in the fields. They were probably owned by the majority of the people, the ones who did not make wills. There were also hens, only listed in two inventories, but likely to have been found in most houses – some of the seventeenth-century inventories list them with the furniture in the living room. Like the cattle, they were kept for home consumption. The occasional roast chicken must have been a welcome dish.

The men and women who left wills were the most prosperous in the village, but for most of the century it seems they were well-off only in comparison with their poorer neighbours. The highest valuation put on anyone's goods until the 1580s was the £24 5s 6d on John Buxton's; six inventories were appraised at under £10 and three between £20 and £30. The villagers with such meagre possessions were still the peasant farmers of earlier centuries, but change was coming. There was an inventory of over £90 in the 1580s and one over £100 in the 1590s, and Shrewsbury Manor rentals of the 1580s show that a few larger land-holdings had been put togther. Hugh Crycllowe (or Chritchlow) had a leasehold of about 100 acres and the two John Lanes, father and son, had about 80 acres of freehold between them, plus a leasehold of about 60 acres. An even more successful entrepreneur was Thomas Westerne, innkeeper and holder of 200 acres at his death in 1621. By this time Westerne was bailiff to the Earl of Kent[9], who had succeeded to the Shrewsbury Manor at the death of Gilbert, the 7th earl, in 1616. Rapid price rises during Elizabeth's reign had benefited these men. Sheep prices had risen from about 1s or 2s before 1550 to about 5s after 1600. Cattle, which had been appraised at less than £1 in the first half of the sixteenth century, ranged around £2 3s after 1600, with a few higher valuations. As the larger land-holders grew richer, especially those who dealt in wool, the small men found it harder and harder to make a living from the land. They found too that when they hired themselves out as labourers that wages were not keeping pace with price rises.

ROGUES, VAGABONDS AND THE DESERVING POOR

The poorest people became one of the foremost concerns of the authorities, those in London as well as the villages. The break up of the manorial system, where every man had his place, the dissolution of the religious houses by Henry VIII, which ended the relief which the Church had traditionally given to the poor and in addition had thrown thousands of monks and monastic servants on to the road, the conversion of arable land to sheep pasture and the failure of wages to keep pace with inflation, had combined during the sixteenth century to create a class of landless poor. Many took to the roads, either to beg or steal or look for work in other parishes. Legislation in Elizabeth's reign, culminating in the Poor Law of 1597, prescribed ferocious punishments for able-bodied men and women judged to be 'rogues and vagabonds', and set up a system of relief for the 'deserving' poor, financed by rates fixed and levied on the land- and house-holders by elected overseers in each parish. The enforcement of these laws was the responsibility of justices of the peace, but the manor-courts did their bit in trying to avoid the cost of maintaining any travelling pauper. This fear of burdening the parish rate payers reinforced the traditional suspicion of the villagers toward outsiders and the fine of 20s levied on Thomas Tyssington by the Shrewsbury court in 1578 'for harboringe suspicious psons' in his inn was an early example of a feature of village life which became more obvious in the following century and persisted through the eighteenth century.

NEW HOUSES

The village houses remained timber-framed for most of the century and thatched for much longer. There was a verdict in the duchy-manor-court in 1615 which suggests that thatch was the almost universal roofing material of the time. Edward Lane was fined 6d for not maintaining his fence at the 'slated house', apparently the only house in the village with a slate roof. Thatched roofs were common in Brassington as late as the nineteenth century, and some survived into the twentieth. A photograph from early in the twentieth century shows the Gate Inn and cottages below it, all thatched.

Delapidated thatched cottage on Hillside Lane, about 1900.

Though timber-framed houses had by then long been replaced by limestone, some of them survived within the new limestone walls; Virginia Cottage, one of five – all now demolished – which stood on Hillside Lane, had a cruck frame, the earliest and simplest of timber frames. The villagers had their own way of making a floor: 'About Brassington, the smallest waste Spar from their mine-hillocks is mixed with a small proportion of quick-lime, tempered together with water, and spread on a floor, and beat down with a flat board once or twice for the first ten days, to prevent its cracking.'[10] One of the five cottages on Hillside Lane was dated 1615; the house on Town Street now known as the Tudor House was built in the same year.

The Tudor House is built of dressed dolomitic limestone but it has interior walls framed in massive oak timbers, the carpenters' marks still showing where the uprights and horizontals were jointed, and the joints fixed by wooden pegs. These walls are either a case of old technology being used in what must have been one of the earliest stone houses, or the original walls of what had been a timber-framed house. Its builder, Thomas Westerne, may have felt that his dignified position as bailiff to the lord of the manor required him to upgrade his house by replacing timber-frame with stone. Westerne's house was an inn. Built a year later is another stone inn, the Gate, positioned at the edge of the village on the other road into it, the Bradbourne Road.

That there was a rebuilding in the late sixteenth and early seventeenth centuries is apparent when inventories are compared. While there is no evidence in the sixteenth-century inventories to identify house types, the small amounts of furniture and equipment must mean that they were small and single-storeyed. John Buxton (1574) lived in greater style than his forebear William Bucstone, who died in 1541, but his household goods would still have taken up little space. His inventory has only four items for household goods. Buxton's house was likely to have had a hall, the main, central living room, with a fireplace (there are fire irons valued at 3s), a parlour (bedroom)

The 'Tudor House', actually built in 1615.

The Olde Gate Inn, built in 1616, with a nineteenth-century extension.

and a buttery. In contrast to this document, nineteen of the thirty wills or inventories proved between 1620 and 1650 name a room or rooms. Naming one room implies others, and these houses, built at the turn of the century, were evidently bigger than the ones they replaced.

Most of the early seventeenth-century houses were still the single-roomed clay-walled structures often, though not always, implied by the word 'cottage' as it was used then. These houses were easily built and the manor-court was alert to new ones appearing illegally on the common land: 'We prsent Alice Topplis dwellinge upon the waste in a Cottage' (1641). Alice was breaking a law of 1589, designed to reduce the nuisance caused by a new class of squatters created by the economic upheaval of the sixteenth century. Large houses were built by the men who had done well out of the changes.

TUDOR HOUSE TO RED LION

In Brassington the outstanding example of this class seems to have been Thomas Westerne. His will, proved in 1622, described him as 'gentleman', a status supported by his flock of 120 sheep and his thirteen oxgangs of land in the village, ploughed by his seven oxen. Westerne had two houses, the larger one having six named rooms and an unspecified number of 'lower chambers' (cellars). This house is the only one named in the inventories of this time which can be identified. It was a copyhold property, its changes of ownership recorded by the manor-court, and, very unusually, it was given a name, New Hall, when Westerne's widow surrendered it to her son Robert in 1636. It was named as New Hall in subsequent transactions in the seventeenth century, until 1680, when it passed to an innkeeper/yeoman from Bonsall called Ralph Marple.[11] When Marple died in 1695, the list of rooms in the inventory included one called the 'Captain's Chamber', identifying it as the same house as the one described in his grandson's will. The grandson, Job, died in 1755, and his appraisers produced a most detailed inventory, including the contents of the Captain's Chamber. The house's history after 1755 is well documented up to the end of the nineteenth century, when the name 'Tudor House' was carved on the datestone which Westerne had set up in 1615, with his and his wife Anne's initials – T.W. and A.W.

Westerne's six named rooms are three first-floor rooms named as 'great chamber', 'little chamber' and 'corne chamber', and, on the ground floor, a parlour, a buttery and the hall. The corne chamber may have been a corner chamber. However, it contained flax, hemp and 'other things' as well as a bed, and may, in fact, have been used as a store room. Bedrooms often doubled as stores at that time. The hall in Westerne's house is almost unique in these inventories – only John Buxton in 1641 had another. The room most often mentioned is 'house', the Derbyshire way of referring to the main living room, still in use in the twentieth century.

The generality of yeomen's houses in Brassington were on a more modest scale than Westerne's. Of the other houses in the wills proved in the first half of the seventeenth century, and likely to have been built at the end of the sixteenth, four have five rooms – John Buxton's (1641), George Wilcocke's (1637), Robert Gratton's (1647) and the house left by Thomas Westerne's son-in-law Henry Trevis (1650). There were two with three rooms and two with two. In addition, there were eight houses which can be assumed to have had several rooms since there were goods listed in their inventories additional to those in a single named room – there must have been other rooms. To these seventeen houses may be added the house of William Greatrax. He was one of the leading yeomen and though his will, proved at the court of the Archbishop of Canterbury, has no inventory, it is unlikely that his house was smaller than the rest of the yeomen's houses.

The full list of named rooms in these early seventeenth-century houses is house (named fourteen times), parlour (eight), buttery (two), hall (two), kitchen chamber (three), backhouse (two), chamber over the house (three), little chamber (two), and store, bedchamber, chamber, great chamber, corne chamber, lower chamber, upper chamber and 'seiled' (panelled) chamber once each. Chambers were upstairs bedrooms, often used as store rooms as well as for sleeping, and parlours were ground-floor bedrooms.

BED AND BOARD

Inside their houses, as the Elizabethan years went by and into the following century, the villagers became more comfortable. The early inventories reveal that members of the Buxton, Charlton and Lawne families had neither stools nor chairs, and only the 'bord' and trestles which preceded tables. They sat on forms. Later members of these families, during the reign of Queen Elizabeth, lived in greater style and enjoyed some comfort. In 1574 John Buxton still ate his meals from a board but had two 'cheyres' as well as a form. He and his family slept in a bed (there were two in the house) on mattresses. In 1562 Thomas Lawne's house had two beds and two stools. By 1582 Hugh Crycllowe, who had been in the wool trade when it prospered and still had a large flock at his death, had eight coverlets for his beds (number unstated), two bolsters, six pairs of sheets and three blankets. Crycllowe's house had other refinements – two table cloths, an 'almerye' or large cupboard with doors, a 'chupborde' without doors and, uniquely for that time in Brassington, a shaving pot and a basin.

By the seventeenth century those villagers who made wills all had tables and chairs. An example is Roger Jackson, described as a miner but whose living came from a combination of mining and farming in a very old tradition in Brassington. His house had two tables and three chairs at his death in 1614, as well as a form and trestles. He also had plenty of storage space for his family's crockery and clothes – two cupboards, a 'dishboard' and a 'dishcradle' for the pots, and five chests for the clothes and linen. Jackson's is the only inventory to mention 'the glasse in the windowes'. It was only during Elizabeth's reign that windows began to be glazed. They had formerly been quite open, though there were often internal wooden shutters which could be drawn across to keep out bad weather. The fact that Jackson's appraisers listed his window glass means that they regarded it as separate from the fabric of the house – it was probably still unusual in 1614. Roger Jackson's house had a form of decoration which had been the only one in earlier times – 'painted cloths' on the walls.

Among the surest indicators of comfort in a house were the fire irons. They give clues to both the heating and the cooking arrangements and, like the bedroom furniture, chairs and tables, became more versatile in the late sixteenth century. John Lane's inventory of 1591 was the first to include a 'landiron', the large grate for supporting burning wood or coal. The full entry reads: 'one Land Yron brandyron & yron sopitte cobards raykens & toungs', valued at 6s 8d. In addition to the large landiron to hold the fire, Lane's fireplace had a pair of cobards which, like so many other things in these documents, are spelled in a dozen different ways (cobirons, cobyrons, goberde, colborte, coborde, cubbord and the rest). These, on either side of the fire, supported the sopitte (spit). The brandiron was a stand or trivet from which cooking pots were suspended over the fire, and the raykens (singular) was a vertical iron band hanging over the fire from which hooks were suspended to take pots and pans. Earlier inventories had listed brandiron and tongs, implying that the fire was burning on the floor of the house. The presence of a raykens in Lane's inventory implies a chimney to hang it from. The earlier inventories suggest fireplaces in an older style, with smoke escaping through a hole in the roof. The late 1500s was the time when houses began to be built with chimneys and John Lane's may have been one of the first in Brassington. His fully-equipped fireplace was typical of later ones among the well-to-do in the village.

THE VILLAGE TRADERS

By the middle of the sixteenth century shops were flourishing in London and developing quickly in other large towns. In the countryside isolation and poor transport forced the villagers to be self-sufficient and in Brassington the people made their own clothes, grew their own food, gathered their own fuel and made their own lamps and candles. Much of this lasted into the twentieth century, but there were nevertheless shops there by the latter half of the sixteenth century and more in the seventeenth. The wills and court rolls between them show that Brassington from the second half of the sixteenth century had at least five alehouses or inns, two bakers, two butchers, two blacksmiths, one mercer and one ironmonger. The mercer, and it is highly likely that there were others like him, sold not only a great variety of cloth, but also many other goods which, unlike ale, bread and clothes, the villagers could not make or grow for themselves.

Anthony Kempe the mercer and his wife Frances died in 1613 and 1617 respectively, and their shop must have been in business in the late 1500s, when their surname first appears in the manor-court rolls. They were a Wirksworth family and Frances's inventory has goods in shops there and in Brassington and Bakewell. This inventory has 338 entries for shop goods, sixty of them in Brassington, and her husband's earlier one even more – 404, listed with no indication of where the shop or shops were. The two inventories have a remarkable variety of the shops' main stock-in-trade, cloth – fustian, linen, scotch cloth, 'sleasie holland' (a flimsy linen cloth), cambrick, silk, taffeta, buckram, velure, velvet, baize, lawn, cotton, buffin (a coarse cloth), canvas, satin and harden (a coarse, hard cloth used for bed clothes and table cloths). There was 'statute lace', binding lace, broad lace, ornamental lace, curled and tufted lace, crewel lace and parchment lace, or lace worked on parchment patterns. The shops sold ribbons, knitted and worsted men's, women's and boys' stockings, silk and brass buttons, silk girdles, pins, combs, thread, string and inkle (tape). The villagers looked for variety in their garters – they were silk, woollen and 'cruell' (crewel or embroidered), as well as simply garters or gartering. There were caps for sale which may have included woollen 'statute caps'. An Act of Parliament passed in 1571 – one of many designed to protect the wool trade – directed that on Sundays and 'holy days' all above six years old should wear 'a cap of wool, knit, thicked and dressed in England'. We can imagine that the women at least preferred to wear something more becoming, and one of ten villagers fined by the Earl of Shrewsbury's court in 1578 for breaking this law was 'Roger Jackson's wyffe', Ann. We know from Jackson's will that Ann was a lady with a mind of her own. He died 'hopeinge that she will not wastfullie and vainelie spend' her inheritance, and 'that she will soe bestowe yt upon necessaries for her selfe as there remaine after her decease some reasonable portion of good.'

The cloth in the Brassington shop in Frances Kempe's inventory was mostly mundane cotton, linen, sacking and scotch cloth. There was half a pound of coloured silk and some 'od lace', and it may be that the village shop kept a less varied stock at all times than the one in the larger community at Wirksworth. However, the Kempes lived at Brassington, and it seems likely that they would have made the full range of their goods available to their neighbours, including some not generally stocked in the village. It must have been a pleasure to visit the shop, perfumed by a quarter of a hundredweight of corande (coriander), two pounds of 'synamon', a pound and a half of mace, half a pound of cloves. For the children of the village the shop had four pounds of 'browne candy', valued by the appraisers at 6s. There were seven pounds of 'woome seeds' to cure the villagers of their intestinal complaints, and to enliven their cooking the shop offered them pepper, nutmegs, aniseed, cloves, ginger and liquorice. There was also starch, brimstone, soap, 'turmericke' (an East Indian powder useful as both a dye and a medicine), and two gallons of 'aquavite', a spirit which could have been brandy. The quantities in the Brassington shop were small – two firkins (a hundredweight) of soap, six pounds of onion seed, twenty yards of linen, twelve pounds of sugar, for instance.

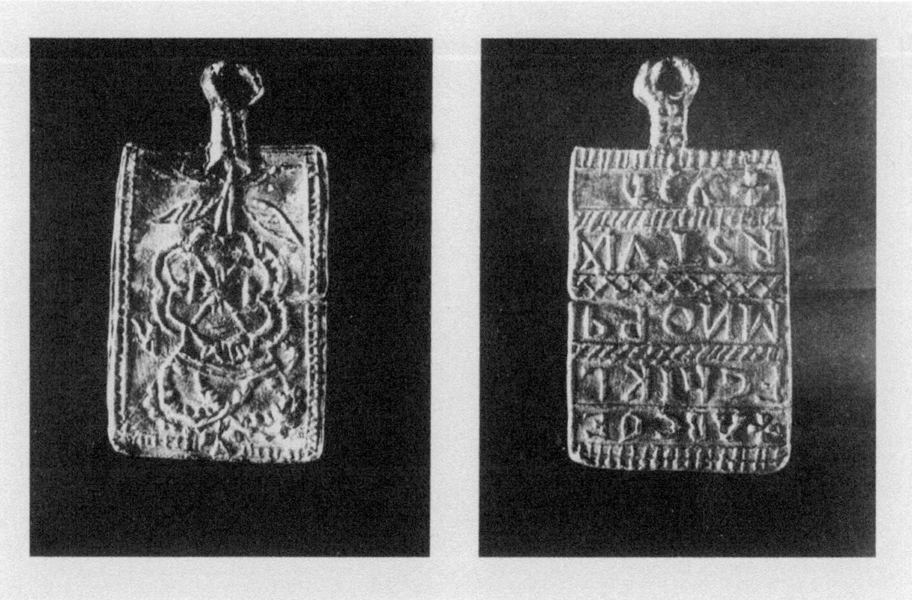

Horn book found in the Tudor House. This photograph originally published in DAJ, 1964.

Silk and lace apart, the Brassington shop's goods had obviously been chosen for a generally workaday set of customers. The presence of eighteen 'cards', used to comb and separate the fibres of hemp and wool, tells us that the villagers made their own rope and spun their own wool. Mrs Kempe's shop made coloured clothing possible. The wives of the village could improve the garments they made with the help of indigo (blue), madder (red) and 'copres' (green). Men and women from the gentlefolk, the Westernes and Buxtons, were presumably the customers for the perfumed 'oile de bay' to annoint their hair. It was also perhaps the same families who bought the 'accidenses & grammers' and the horn books – a horn book was an alphabet and numbers mounted on an object shaped like an oblong table tennis bat and usually covered in transparent horn for protection. It was intended to make the alphabet and numbers familiar to a child by being ever-present – it was often hung from a child's belt by its handle. However, in both education and fashionable dress and perfume the customers probably, by the end of the sixteenth century, included the flourishing yeoman families – the Lanes, Allsops, Knowleses, Crycllowes. They too would have been able to afford sugar, a luxury at 1⁄2d a pound, and the only ready-made garments in the Brassington shop, four pairs of stockings at 1⁄3d each.

Some of the products listed in the two Kempe inventories which are not mentioned among the Brassington goods but which must have been sold there, since they are the everyday things which would sell as well in the village as at Wirksworth, are vinegar, treacle, honey, liquorice, turpentine, sugar candy, garters, caps and abacuses. Tobacco, only recently brought to England by Sir Walter Raleigh, may not yet have caught on at Brassington, and one commodity does pose a problem – gunpowder. Firearms are not mentioned in the Brassington inventories before German Buxtons's two 'fowling peeaces' in 1686, and the use of gunpowder for blasting in the lead mines is believed to have come late in the seventeenth century. Perhaps the sale of gunpowder was restricted to the larger population of the market town, where there probably were guns. Much of the cloth not listed in the Brassington shop must have been sold there – fustian, canvas and buckram were the cheapest working cloth. Perhaps the most expensive was calico at 1s 8d a yard and taffeta, very costly indeed at either 6s 10d or 8s a yard, may only have been available at the Wirksworth shop. Fustian varied from around 1s 2d a yard; canvas was about 1s and buckram, a coarse linen or cotton fabric, was valued at about 8d a yard.

Every farming community needed a blacksmith. The growing number of horses in the late sixteenth and seventeenth centuries was increasing the need for iron shoes for draught animals, since oxen were shod only on the front feet and needed shoeing less frequently. Iron shares were needed for the wooden ploughs and iron tips for the spikes of the wooden harrows. The villagers needed nails, door handles, hinges, spades, forks, chains and a score of items which a blacksmith could make for them. Working in the village during the last quarter of the sixteenth century was Henry Spencer, one of a family who were blacksmiths for at least three generations. When Henry died in 1632 he left a will and an inventory. The appraiser's handwriting is so bad that the inventory is almost indecipherable, but the contents of the blacksmith's shop can be made out. A manor-court entry of 1628 described the 'smythes shoppe' as containing one bay. This 12ft-long smithy, half the size of Henry Spencer's cottage, had a hammer, a shoeing hammer, a bellows, three pairs of tongs, a pair of pincers and a 'punger' (possibly a punch), showing that the English blacksmith's tools changed little over the centuries. A near contemporary of Henry Spencer's, Henry Allsop, had the same equipment, plus a 'stiddie' – an anvil. We can only assume that the appraiser with the bad handwriting was also careless, since Spencer must have had an anvil too.

One of Spencer's tools needs an explanation. This was a 'nugar' or borer – he had two. This word, also spelled 'noger' or 'nauger', was the original word which later became 'auger' in standard English, and survived in lead mining after it had disappeared even from standard Derbyshire dialect. Henry Spencer may have made his two nugars for a customer, perhaps a joiner or a miner, or he may have used them in his own trade. Nugars were used in mining as part of the process of hammering in wedges to split ore-bearing rock. They were also used to bore holes to take gunpowder for blasting and the presence of two of them in Henry Spencer's smithy may be a second piece of evidence to add to the gunpowder in the Kempe's shop, pointing to its use in mining earlier than has been thought.

Bread was baked and beer brewed in many homes. Home brewing is obvious from the mentions in inventories of brewing equipment. However, the regular fines imposed by the manor-court for breaking the Assize of Ale reveal that there were professionals in the village. As we have seen, there were alehouses at least as early as the beginning of the fourteenth century. By the sixteenth there is more evidence: according to a survey conducted in 1577[12] there were four at that time in Brassington. The survey was made by Sir Francis Leake, chief justice of the peace in Derbyshire, and was part of a national survey designed as a preliminary to raising revenue from ale for the reconstruction of Dover Harbour. It was sent to the Privy Council in London. The four alehouses in Brassington were kept by Humphrey Lund (Lane), William Walton, Thomas Tissington and widow Tissington, and in the same year the three men were fined 4d each for breaking the Assize of Ale – the manor had raised money from ale long before the Privy Council thought of it. Alehouse keeping seems to have been a family tradition in the case of the sixteenth-century Tissingtons. In 1542, John Tissington had paid his 4d to the manor. The Derbyshire survey says of the alehouses that 'many are very poore', but the Brassington publicans included some of the more prosperous men of the village. The leading 'gentleman' at the end of the century, Thomas Westerne, was an innkeeper and the three men named in the 1577 survey were all prominent in the village. All served at times on the manor-court jury and the court rolls show that Walton occupied land and a croft as well as his alehouse. Tissington and Lane were yeoman farmers.

A few incautious words after drinks in William Walton's alehouse in May 1578 led to one traveller, a Lancashire carrier, being hauled before the local JP, Anthony Gell.[13] Thomas Whalley and Ralphe Wennesley, 'ii cariers beynge Lankyshre men', made the mistake of pulling Roger Skynner's leg by joking about his master, who was presumably the Earl of Shrewsbury since Skynner was one of his tenants. Whalley complimented Skynner – 'this man dothe belonge to the beste man in Derbyshyre' – and Wennesley played his part in the game by contradicting him – 'he thought he belonged to the worste man in Derbyshyre'. According to Skynner's statement, Wennesley amplified this by saying that his master was the most covetous man in the county, though Wennesley insisted that he had merely said 'hardest'. Whalley and Wennesley had chosen the wrong man for their joke. Skynner, who was often in trouble himself for breaking the farming rules, was the one who tipped off the manor-court about his neighbours breaking the law about wearing woollen caps. Skynner's and Wennesley's statements survive in the Gell papers in the Derbyshire Record Office, though sadly without Anthony Gell's verdict.

Alehouses, being simply houses whose occupants sold ale or beer, were as elaborately or as simply furnished as the host could afford. Thomas Westerne's house was furnished grandly for the time. One of his contemporaries in the trade, a later member of the Walton family, left an inventory which reveals that his customers sat on forms at a large table and shared the room with '2 piggs and 4 henns'. What his house had in common with Westerne's was a fire, probably as welcoming to his customers as his ale. The alehouses were valuable meeting places, unofficial and free-and-easy. They gave credit when money was scarce, and the debts, amounting to £6 3s 4d, still owed at the death of another Walton alehouse keeper, were an occupational hazard.

Alehouses were also refuges for the travelling poor. An alehouse would shelter a man for the night when no one else would, for fear of being brought to court, manor or assize for harbouring an 'incomer'. Thomas Tissington ran the risk at his alehouse in 1578 and was caught, as we have seen. There were alehouses in other places which supplied forgeries of the passes which travellers had needed since 1531. These were signed by a JP and certified where a traveller was headed and why he was travelling. Without his pass he could be imprisoned and flogged. Perhaps one of the Brassington alehouses gave this service.

SOLDIERING

While the villagers' lives were affected by the administrative changes brought in by the Tudor monarchs, and by the economic changes which by the end of the reign of Queen Elizabeth were demolishing the old communal style of farming, they were also touched by the wars and preparations for war against foreign enemies. Three of the sixteenth-century Derbyshire muster rolls, or lists of 'able men' available for army service, mention men from Brassington. The first is the muster ordered by Henry VIII in 1538.[14] On this occasion Brassington contributed twenty-one names to the Derbyshire muster, a very high number considering that there were probably fewer than 100 adult men in the village. The men are grouped by their military specialities. John and William Charlton are described as mounted armoured archers, a strange category for a sixteenth-century English army. Equally unlikely is the combination of horse, armour and 'bill' ascribed to James Wright, Thomas Daken and George Buxton. A bill was a spear with a hook at right-angles to the spear point – a standard infantry weapon of the time. The armour and horses of these men suggest that they were among the better-off in the village. Bows and bills were used by foot soldiers, and it sounds as though the five horsemen had listed every weapon in their houses. There were two archers, Thomas Tissington and Richard Gratton, and a list of fourteen 'bilmen'.

Twenty years later Queen Elizabeth called a muster.[15] To the fairly modest Derbyshire force Brassington gave 'harnes in redynes for one bill man' and five 'able men wt out harnes'. This, while a large drop from the size of the earlier muster, was still likely to have been about 5 percent of the men of fighting age in the village. In 1587[16] a muster was called to counter the threat of invasion by Spain. This roll has the soldiers' names and from Brassington came Thomas Wallwyn, armed with breastplate and bill, putting his fighting spirit to better use than when he had been caught brawling in the village nine years before.

THE REFORMATION

Henry VIII's break with the Church of Rome swept away what must have seemed to the villagers to have been a fixed and timeless part of their daily lives. Brassington's church, governed by distant Dunstable Priory, had stood on the hillside for over four centuries. The villagers' breaks from work were taken on the Church's feastdays – Easter, Whitsuntide and Christmas. Every unfolding year was marked out by saints' days: 'noe person carry or recarry thorowe ye medowes before Mighelmasse', 'no person doe putt into the fallowe any lambes before Saincte Barnabyes day next'. Transactions agreed in the manor-court had often given saints' days for their execution and the custom continued: on 12 August 1589, for instance, land in the occupation of Henry Myddleton was transferred to Thomas Lane, with effect '*ad festi Sci Johes Baptisti et Sci Martin Episcopi*', or from John the Baptist's (29 August) and Saint Martin's feast day (11 November). A favourite date for land transfers in the sixteenth century was '*a festo annunciationis bte Marie Virginis*' – at the feast of the Blessed Virgin Mary, or Lady Day (25 March, New Year's Day in the Julian calendar of the time). The villagers were baptised in the church and buried either there or in consecrated ground outside it. In 1545, Thomas Lawne bequeathed 'my soule to almyghtie God and to his blessed mother... my bodye to be buried in ye churche of Seynt James of Brassington.' He was no doubt following a custom which was several centuries old by then.

The churchyard, from Hillside Lane.

House next to the churchyard, said to be a former vicarage. It is perhaps the sixteenth-century 'preste hows'.

There is a payment of 13s 4d by Humphrey Alcocke, 'ye curat there for ye church yard', in a Shrewsbury Manor rental of 1589. Since it had not appeared in earlier rentals, including that for 1588, it seems that it was only in 1589 that the church acquired the burial ground which now surrounds it. Wills earlier than 1589 provide for burial in the church, later ones either in the church or the churchyard. (German Buxton in 1620 not only directed that his body be buried in the church, but specified that it should lie in the place where he had worshipped during his life – 'in the seate or forme wch I usual-lie sit in').

Thomas Lawne died after Henry VIII's quarrel with the Pope but before Elizabeth had imposed a new Church on the people. Her settlement of the religious course which England was to follow came after a swing of the pendulum to Protestantism during Edward VI's brief reign and a swing the other way, back to Roman Catholicism, under his sister Mary. Since worshipping in the wrong way could mean a particularly nasty death by burning, the villagers probably swung with the times. There was a great deal of enthusiastic plundering and smashing of 'popish' Church property and decoration during Edward's time, and Church goods were often sold, the proceeds being used for the upkeep of the fabric of the churches, the repair of bridges and highways, the purchase of 'harness' and the 'setting forth of soldiers', and alms for the poor.[17] These were the official uses – in fact many treasures were simply stolen by the local gentry. There is no evidence as to what happened to the pre-Reformation property of Brassington church but we do have a list of what was left after it, as an inventory was taken on 30 September 1552.

The church possessed a silver gilt chalice, a vestment for the parson 'with albe & amysse' (a white surplice-like garment and a hood respectively), a surplice, a towel, alter cloths, a 'corporas case' and a 'corporas cloth' (a eucharistic vestment and a cloth on which eucharistic vessels were placed and with which they were covered when not being used), one 'sanctus bell in the stepull', and four other bells, one a handbell. In Elizabeth's reign measures were taken to stop further attacks on churches by over-zealous Protestants.

The 1589 and 1590 Shrewsbury Manor rentals also have an entry for Thomas Alcocke and Robert Briddon paying 6s 8d for 'a cottage called the Preste Hows', perhaps a house which would now be called the vicarage. A list of the Lichfield Diocese clergy for 1602[18] reveals that the curate at Brassington was neither qualified for his job nor well paid for doing it. Edward West was curate at both Brassington and Ballidon. He had neither a university degree nor a licence to preach and the annual income of Brassington church was £8. Later curates included the lazy Roger Gretton, described in 1614 as 'sayeth not service in due time, leaveth the church unserved', and the alarming Edward Hollingshead (1636-39) 'who (being as we verily believe distracted), said he lived incontinently with one Jane Bount and further said he was dead and buried and that he was the Messiah and that whoever believed not in him was damned'.[19]

REFERENCES

1 Riden, 1978
2 DRO QSZ Book 4
3 NA DL28/26/2
4 NA DL43/1/19
5 DRO D258 s/ll
6 NA DL43/1/19
7 NA E317 Derb 28
8 MLSL Lambeth Palace Library 697 f137
9 SCA 128
10 Farey, Vol. 2, 1813
11 DRO D161 B/6/52
12 Hunt, 1879
13 DRO D258 44/2k
14 NA E101 59/7
15 NA SP 12/3
16 Carrington, 1895
17 Walcott, 1870-71
18 Cox, 1884
19 Clark, 1984

CHAPTER 4

EARL GEORGE

SHREWSBURY ESTATE MANAGEMENT

The duchy-manor tenants were governed by a remote and vast bureaucracy, stretching up from the village reeve to the bailiff for the Honour of Tutbury, and the 'receiver' at Tutbury, and on to the high officials of the Duchy Council, including the receiver-general, or chief financial officer, and the chancellor himself.[1] In contrast, their neighbours during the sixteenth century were under the personal scrutiny of George, 6th Earl of Shrewsbury. Brassington was only one of Earl George's many manors, but he was a vigorous entrepreneur and his activities had direct effects on his tenants there. He was a national figure, a favourite of Queen Elizabeth. For this reason, and because he was one of the richest men in England, he was picked to carry out the task for which he is best remembered in Derbyshire, the imprisonment of Mary, Queen of Scots. He incurred great expense and, no doubt, permanent anxiety, in keeping Mary in appropriately regal style while making sure that she was prevented from causing trouble to the queen. Earl George had a second heavy call on his fortune, his expensive wife Elizabeth – Bess of Hardwick.

The earl supervised his estates closely, and was heavily involved in the lead industry. In both of these activities, remote from them as he was, he changed the lives of the yeomen, husbandmen and miners of Brassington. A clear picture of the way the earl operated is given by his accounts and correspondence. Letters from his bailiff, William Dickenson, show how close an eye the earl kept on his estates. In one letter, dated in 1580, Dickenson tells the earl how at Handsworth he discovered gleaners among the stacks 'wch are not yet ledd in' and drove them off.[2] He sends news of one of the earl's tenants, sick with something Dickenson calls 'mearesands': 'it begane at the toppe of his Thigh or hangbone & so went downwards almost to his knee & then turned up agayne & so almost about his Bellye.' The patient was being cared for by 'the ii wenches there', and his gruesome complaint was being cured by an equally gruesome remedy – 'catts blood'. This sounds like the witches' brew in *Macbeth* – 'eye of newt' and the rest – but must have been the name of a remedy less exotic than the blood of an actual cat.

The letter is dated 12 August, and Dickenson reports that the hay is in at Burton and Breerly – forty-seven loads and sixty loads respectively. The barley is 'shorne' – 'viiixx thraves' (eight score thraves of twelve sheaves to the thrave) – and 'as Steele thinks iiixx thraves or above still to come.' Dickenson relays requests from local officials to pay wages, sends news of land lettings and receipts, and reports on the performance of the earl's servants. A letter of 1575[3] reports that Humphrey Fullwood, the tenant of Gibbe Fields in Hognaston parish, was clearing the ground of thorns and putting it to 'tiledge'. In Dickenson's opinion this was harmful to the land and he had ordered Fullwood to stop his land-clearing while he sought the earl's opinion. With his retinue Dickenson visited every manor where the earl was lord, collecting in his master's dues. Dickenson's record of these receipts has survived, every page countersigned by the earl himself.[4]

There are hardly any payments in these accounts. Those for 1589 at Brassington include 4d paid to 'the wyffe' for providing his party with meat and drink, presumably at an inn or alehouse, and a further 4d to 'the fellowes' for attending to the horses. These payments are set against £88 12s 2d received in rents from tenants, a figure which for two years had included £14 6s 8d for certain pieces of enclosed land.

Earl George ran his estates in a distinctly modem, capitalistic manner. Apart from the nominal 'chief rents' amounting to £1 6s for about 100 acres of land, for which a number of villagers held the freehold, his tenants held their land on a 'farm' or leasehold basis. The twelfth-century grant of the lands which eventually became the Shrewsbury Manor must have been of the freehold, and while there can have been no advantage to the lord in leasing this land to the villagers during the stagnant years of the fifteenth century, by Earl George's time farming had become profitable, and he made the most of his freedom from the restrictions which copyhold imposed on the duchy of Lancaster. A comparison of the Shrewsbury Manor rents during the 1580s with the figures for 1619[5] shows a rise of almost 100 per cent in the total income. It is likely that the leases were renegotiated as they expired, with a higher rent the usual result. The earl's income from his Brassington Manor was also rising as a result of a different development. There were 'closes' (enclosed fields) in Brassington before 1580 – for example, the 1575 letter from William Dickenson refers to 60 acres of enclosed land at Brassington let at 3s an acre. Although most of the land was still open field, Earl George was making decisive moves toward the enclosing of the village's farmland.

Sheep farming had been the main reason for earlier enclosures, but by the closing years of the sixteenth century lords and tenants were enclosing for more efficient arable farming, and for the opportunity to farm individually, without the need to stick to a village pattern. The open fields hampered crop growing in a number of ways. They could only be worked as efficiently as the least efficient of the villagers managed his strips since everyone had to grow the same crop as his neighbours and sow and reap it at the same time. The strips were manured only by the sheep which were folded on them for half of the year and by the sheep, cattle, pigs and horses which grazed them during the year when they were left fallow – they were in consequence poor croppers. The scattered strips, scattered in the first place to make sure that good and poor ground was fairly distributed, were a further hindrance to efficient farming – many were difficult for a farmer to reach without crossing someone else's land. Nevertheless, the complicated tenure systems, combining copyhold bound by the custom of the manor, leasehold for specified periods, and freehold, made change difficult and therefore slow. The small copyholders, fearing the loss of grazing right which all men had on the open fields at prescribed times of the year, were always likely to oppose enclosure schemes.

A document in the Shrewsbury papers at Sheffield[6] reveals that in the 1580s Earl George was 'treating about Exchange of some freehold & copyhold lands with some tenants at Brassington for the better Division of the Lordship.' These transactions had the effect seen in the 1588 rental for the manor, which includes entries for both open-field holdings and enclosed fields. In the following entry, John Harrison's holdings on the open fields are charged at 20s per year: 'Jo harryson for v pcells of land in the Towne fields called parr furlong lowe cliffe Balke and the short brech & the buckleather pn xxs.'

The fields called Lowcliffe Back are shown on a nineteenth-century parish map. They are on the south-west of the village and amount to just over 9.5 acres. Buck(s) leather is in the same area and consists of 3.75 acres. Parr (Spar, Spur) Furlong is also in the south-west and covers 10 acres and while Short Brech is not named in the 1835 survey which identifies the fields on the parish map, it can be taken that it was in the same area, apparently one of the original open fields – Town Field. Other entries show the origin of some of the enclosed fields:

Aerial view of fields and roads. Those to the south and south-west have hardly changed from when they were formed from collections of adjacent strips, and are quite different from the rectangular fields in the north marked out by the enclosure commissioners in the early nineteenth century. The fields below Rainster Rocks, near the top left of the picture, part of the 'waste' enclosed in 1808, have the strip pattern revealing that they were farmed in medieval times and then abandoned.

Jo Lane for a close pceld of the demesne nowe enclosed p an ixL iis
Rich Knowles & Tho Buxton for another close nowe enclosed pceld of the demesne there iiiL xiiis iiiid
Jo Wright for a peece of a close callyd the pease lands pcelld of the demesne lands and for another close xxvis viiid

The two closes held by Lane, Knowles and Buxton may have been the ones later listed among Richard Buxton's possessions when he made his will in 1631. He named them as the meanes, 'late pcell of the freehold possessions or inheritance of the Right Honourable Gilbert Earl of Shrewsbury.' The name 'meanes' or 'meaney', clearly the villagers' way of saying demesne, is still in use as a field name in Brassington. The Pease Lands, amounting to slightly less than 8 acres, are to the south of the village, half a mile from Harrison's holding and a good illustration of how scattered one manor's territory could be. What the earl and his tenants did was to exchange strips so that the land of each party was held in adjacent strips, suitable for fencing off from the rest of the arable land. Having created his fields, Earl George let out that part of his demesne land for rent to village farmers. These enclosures amounted to 94 acres or 19 per cent of the 497 acres of the Shrewsbury Manor. John Lane's close seems from the rent of £9 2s 8d to have been a field of 60 acres, calculating from the rent of 3s per acre mentioned by William Dickenson and suggested by other figures among the rentals. This large field is untypical of the pattern at Brassington, where the presence of small freeholders and copyholders made it difficult for Shrewsbury, the duchy or their successors to create large fields – the small man's contribution to an enclosure agreement was rewarded by fields as large as his few strips. Lane's large field was presumably used as grazing for sheep since he had a flock of seventy-four ewes and thirty-one lambs at his death. William Dickenson's letter of 1575 notes another source of income for Earl George. Since he owned the freehold of his manor, he had the grazing right on the commons which went with it, and was letting this out at an annual rent of 3s per beastgate.

LEAD BUSINESS

The second great source of income from his Derbyshire estates was lead. The ore-fields had been unproductive in the early part of the sixteenth century, largely because the price of lead fell when the market was depressed by a glut from the monasteries and abbeys dismantled after Henry VIII's dissolution of the religious houses. The size of the glut can be gauged from one sale, recorded in a letter in the Shrewsbury papers in Lambeth Palace Library.[7] This refers to the sale of 3,000 fother (a fother or fodder was about a ton) at £4 6s 8d a fother by the Court of Augmentations, the body set up to administer the sale of the religious property. The same letter ordered that the lead remaining on the Abbey of St Mary in York and on the monastery at Dunstable be removed, in order to fulfil a bargain made with a Genoese merchant called Ancelyn Salvago – there was a strong export trade. The depression was ended by two developments. One was a revival in demand caused by the building of new mansions. Houses such as Hardwick Hall consumed great amounts of lead sheet for their roofs, lead piping and guttering and water tanks, and there were miles of lead in the windows. The second was the introduction of the smelting furnace for which William Humphrey was granted a letters patent in 1565.[8] This furnace used bellows and was able to extract much more lead from the ore than had been possible with the ancient 'bole hill' method, which consisted of alternate layers of logs and lead ore and which relied on steady wind power. The earl and Humphrey seem to have collaborated, as letter from Humhrey reports his petition to the queen in 1578 for the granting of a monopoly to his new process[9], and a letter from John Manners three weeks later[10] asks that Humphrey's 'commission' should not prevent him from using his own 'footblast' in smelting, as he hoped to conclude an agreement which depended on this new method.

The village miners sold their ore to the smelters, who in the second half of the sixteenth century were the agents of Earl George, owner of the most productive mills in the county. The real fortunes were always made by the smelters and, since setting up a smelter and organising the sale of the smelted lead required capital and contacts with consumers, there are few instances of miners smelting their own ore. That they were aware of the situation and would have liked to have changed it if they could, appears in a letter sent to the earl by two of his ore buyers in 1585.[11] They wrote from Ashford and reported that they had bought ore there and had then gone to Over Haddon. Here the miners, who were producing the best ore in the Peak, had refused to sell their 100 loads ready for smelting. The buyers awaited the earl's orders and commented, 'now everie miner that haieth a good grove wilbe a burner [smelter] & the lord of the Feld shall not have their ourre but by force.' The earl's reply has not survived, but he is not likely to have been deterred by the need to use force, and however the Over Haddon confrontation was resolved he continued, to smelt the ore from his manors at his mills at Totley, Cromford, Crich, and Wooley.[12]

Another letter is evidence that the earl sold some of his lead abroad.[13] Written from the ship *Talbot* in 1579 it discusses the disposal of pigs of lead and reports that some remain unsold at Rouen. Other letters name Brassington as an important source of ore. The manager at Earl George's Higham mill reported in 1578 that he was buying about fifty-four loads a week from Taddington and Brassington (there were about four loads to a ton). Although there is a letter in the Shrewsbury collection at the College of Arms stating that the mines in Derbyshire were in decline, those in Brassington seem to have remained profitable until after the earl's death in 1590, as a letter to his successor, Gilbert, refers to the 'orr that is now gotten in Brassington lordship... that is verie good orr and great stor'.[14] Earl George, in fact, had known enough about the profits to be made from the mines at Brassington and the other liberties in the Wirksworth wapentake to take out a twenty-one-year lease on the duties paid by the miners.[15]

Fifty years after Earl George's death his family's long tenure of the manor of Brassington came to an end. His son Gilbert died in 1616 and the manor was then held jointly by Gilbert's three daughters, married to the earls of Kent, Arundel and Pembroke. These successors, still perhaps feeling the effects of the extravagant spending which had been the other side of the coin during Earl George's money-spinning days, sold out to William Savile, gentleman, of Bakewell.[16] Savile seems to have made a good bargain. He paid the Earl of Kent £950 for his third of the manor in 1639 and in the following year completed the bargain by paying the earl of Pembroke £1,300 for the rest. With these sales and the arrival of local gentry to take the place of the great noblemen, a link with the village's medieval past was broken.

REFERENCES

1 Blanchard, 1971
2 SCA Bacon/Frank Collection, 2/44
3 MLSL Lambeth Palace Library 697 f137
4 SCA ACM 8114, 8117, 8118
5 SCA A128
6 SCA WD924
7 Bill, 1965 (MS 707 f161)
8 Ford, 2000
9 Bill, 1965 (MS 697 f79)
10 Bill, 1965 (MS 697 f98)
11 SCA Talbot Correspondance 2/78
12 SCA Talbot Correspondance 2/79
13 & 14 Kiernan, 1989
15 Bill, 1965 (MS 705 f148)
16 MLSL BM Add MS 6690 ff1-3

CHAPTER 5

SQUIRES, YEOMEN AND THE POORER SORT

THE VILLAGE YEOMEN

Who were the yeomen who held the land in the two manors and were the sometimes reluctant reeves, constables and overseers? There are fourteen family names which were listed either in the Shrewsbury Manor rentals or the duchy-manor juries in the 1580s and which were still there in the rentals of the two manors in 1639 and 1640 respectively. Taking a longer scale, there were also fourteen names mentioned in the duchy-manor rolls of the 1560s which were included in the manor's suite roll in 1664. Nine names appear in both of these lists – Allsop, Buxton, Charlton, Gratton, Knowles, Lane, Tissington, Torr and Wilcocke. Of these family names all but Torr appear in manor-court juries in both the 1560s and the 1660s. These families were all living in Brassington in 1500 and they provided sixteen of the thirty-two men who formed the manor-court juries of the 1660s. Clearly the village had a nucleus of farming families who survived all the economic, political and religious changes of the Tudor and Stuart centuries.

The earliest statistical snapshots which are useful in trying to establish who the haves and who the have-nots were in Brassington are the surviving returns made for the Hearth Tax. The earlier tax returns, the Lay Subsidies, caught very few people in Brassington and show only that members of the Lane, Wilcocke, Allsop and a few other families were the richest in the village. The Hearth Tax was first levied in 1662. It was a most hated burden and was abolished as one of the first measures which William III took to win popularity when he became king in 1688. The surviving returns for Brassington, which was grouped with Aldwark, are for 1662[1], 1664[2], 1670[3] and 1672[4]. A tax of 2s was raised on each fireplace in a house and returns were made in 1662 to the High Constable in Wirksworth by the Petty Constable, who was the manorial official. The High Constable was Richard Buxton and Brassington's constable then was Henry Trevis. Buxton's returns were forwarded to the Sheriff of Derbyshire, Sir William Boothby, for onward transmission to the government. This system did not work well enough for the government and later returns were made by the justices of the peace.

There is a rough-and-ready guide in the number of hearths for which a man was taxed to the size of his house and therefore to his wealth and social standing. A further guide is given by the fact that householders were exempt from the tax if their houses were valued at less than 20s per annum or if they were receiving parish relief. The 1664 return lists those exempt. They were about half of the householders listed, and if we add the thirty-nine villagers taxed on one hearth to the forty-six judged too poor to pay anything, we have a village of very poor people, with a few prosperous farmers and gentry. The population of the village in the latter half of the seventeenth century was about 350, including children.[5] There were nine villagers paying tax on two or more hearths in 1664. Five of them were from the long-established yeoman families – German Buxton, Edward Lane, Richard Goodwin, Richard Knowles and Ralph Brunt.

RALPH BRUNT AND THE TUDOR HOUSE

Brunt was an innkeeper, living in a house identified when he bought it from Robert Westerne in 1655 as New Hall, the house which Westerne had inherited from his mother, Anne, in 1636. Brunt paid tax on three hearths, a figure which had risen to five by the time of the 1670 assessment. Brunt's innkeeping seems to have been a precarious business. He mortgaged New Hall to William Savile in 1660, redeemed it in 1661 and mortgaged it for a second time in 1668 to a Carsington farmer called William Taylor. The manor-court records are missing between 1673 and 1700, but Brunt's fortunes are revealed in a document recording an arbitration on a legal struggle in 1680 between him and Taylor for possession of New Hall.[6] Taylor appears to have foreclosed and then sold the inn to Ralph Marple. Brunt contested this and the arbitrators instructed Taylor to pay Brunt 25s and Brunt to surrender New Hall to Marple at the next meeting of the manor-court. It is this document which links Thomas Westerne's New Hall with today's Tudor House, and confirms the identities of 'T.W.' and 'A.W.' on the house's datestone.

The other four paying tax on two or more hearths were gentlemen – Robert Westerne, Henry Trevis, William Blackwall and Richard Buxton. Westerne, Trevis and Blackwall were the son, grandson and son-in-law respectively of Thomas Westerne. The younger Westerne had inherited his father's large land-holding in Brassington; Blackwall was an innkeeper and a member of a family prominent in the lead-mining industry for over a century, and he and Trevis' father, another Henry, were barmasters (local officials) for Brassington. Land, beer and lead was a most profitable combination of interests for the Westerne family during the seventeenth century. The highest number of hearths declared in a Brassington house in 1664 was in the one owned, though not then occupied, by Richard Buxton, who was probably the man then carrying out the duties of High Constable for the Wirksworth wapentake. This house had five hearths. Later Hearth Tax returns added George Wilcocke and Robert Allsop of the old families, Daniel Bagshawe, ironmonger, and William Dodds, innkeeper, to the number of housholders with two or more hearths. Dodds had four in 1672, and it was at his inn that the manor-court met when Ralph Brunt surrendered New Hall to Ralph Marple.

The rentals for the two manors had identified the prosperous farmers in 1640.[7] There were ten men holding more than two oxgangs, or 30 acres, of land, a figure broadly in line with the numbers of men paying tax on two or more hearths in the 1660s and 1670s. The ten land-holders, about 10 per cent of the village's householders, held forty oxgangs – 71 per cent of the fifty-two oxgangs of the two manors. The Hearth Tax return for 1672 shows a slight rise over the ten years, perhaps an indication of a rise in village incomes, but the overall picture among the taxpayers is not significantly different. A later taxation document, the assessment made in 1677 for raising money to buy warships[8], lists fifty taxpayers. The assessment was to raise £6 9s 5d quarterly, the taxpayers 'being equally assessed', which in this case meant according to ability to pay. The list includes three absentee land-holders, the Duke of Rutland, Sir John Gell of Hopton and Edward Pegge, one of the lords of the manor. The highest payments were from Richard Buxton and from the widow of the yeoman Robert Allsop of Lydgate – 9s 4d and 9s 2d respectively. Thirty-eight of the others were assessed at less than 3s. Of those who paid more, only Buxton and one other, 'Mr' William Shallcrosse, were ranked as gentlemen, and Shallcrosse did not live in the village. The rest were yeomen, husbandmen, miners, shopkeepers and craftsmen. Brassington's population was a working community.

THE BUXTONS AND LANES

At the top of the village social tree by the latter part of the seventeenth century and prominent in village life at least as early as the beginning of the fifteenth, were the Buxtons. Until the seventeenth century the family members were described as yeomen, and there is no evidence from their wills or inventories that they were rich. However, the wills of German Buxton (1686) and of John (1699), who was lord of the manor at the time of his death, are the wills of quite wealthy men. The Buxton lords of the manor were descended from the Richard who died in 1632: his son John bought part of Pegge's manor[9] and John's brother Henry and son John were successively lords of the manor. Other Buxton lords of the manor, George and German, were probably members of a related family.

One of the bequests in Richard Buxton's will in 1631 was of land formerly held by Frances Ashmore. She was his daughter by his first wife, Margaret Palfreyman, and Frances, as the heiress of her maternal grandfather, George Palfreyman, had been the object of what seems to have been a bitter dispute between Buxton and a Carsington yeoman called Anthony Ashmore.[10] Ashmore, no doubt with the heiress's great expectations in mind, eloped with Frances. Buxton, even more probably for mercenary reasons, laid charges of abduction against him in the Star Chamber. The quarrel was settled by the mediation of Francis Fitzherbert of Tissington in 1614. Ashmore ended up the clear winner at that time, keeping both the girl and the land, though Buxton seems to have got his own way in the end, regaining the Palfreyman inheritance after Frances's early death. Richard Buxton's will reveals a connection with the town of Buxton in a bequest of land there to his second son Henry, and his grandson Richard: it seems that when the king's herald was checking on the family's right to a coat of arms in 1662[11], he claimed that the elder Richard came from Buxton. However, a second grandson, John, son of Henry, by then living in Youlgreave, described his grandfather as being of Brassington and 'possessor also of lands in Buxton'. Richard had, in fact, made his background clear in his will, where he instructed that his body was 'to be buried in the Church of Brassington aforesaid where my Ancestors were buried.'

Two of the freeholders in the Shrewsbury Manor in the 1580s were John Lane, father and son. The Lanes and Buxtons often appear as witnesses or appraisers in each others' wills and inventories and were obviously social equals. Richard Buxton's second wife, the mother of all his children except Frances, was Elizabeth Lane, probably the daughter of the John Lane, who died in 1597. When a later Lane, Edward, died in 1666, his will revealed that he was living in a seven-roomed house, a house which his widow's detailed inventory eight years later showed was exceptionally lavishly furnished.

THE WESTERNES

Rivalling the Lanes and Buxtons in land and possessions were the Westernes. Whether or not Thomas Westerne was descended from an old Brassington family, he was by the end of the sixteenth century playing a major part in village life. He was one of the men often called in to help with his neighbours' wills – he was appraiser on five occasions, and a witness of Frances Kempe's will in 1617. He was also a creditor, with George Harrison, for instance, owing 10s to 'goodman Westren' at his death in 1614. Westerne, bailiff to the Earl of Kent, left thirteen oxgangs of land in Brassington and more in Carsington when he made his will, which is the only Brassington will at Lichfield Joint Record Office written on parchment, presumably chosen to suit his status. There were more than seven rooms in Westerne's inn – more than were listed in any Brassington house before the end of the seventeenth century – and his inventory valuation, £277 17s 4d, was more than £140 higher than the next highest (Anthony Lane's) apart from the special case of the stock-in-trade of the Kempes, until 1652, when German Buxton's appraisers came to a higher figure.

By 1640, Westerne's son Robert held the 'capital messuage farm or tenement of nine oxgangs of land meadow or pasture' in Savile's manor. The term 'capital messuage' signifies a large house. The Westerne family connections included Thomas's in-laws, the

Newtons of Okerthorpe, who were gentry. His sister was married to Ralph Gell of Carsington[12], Robert's wife was a Buxton, and Robert's sister Helen married Henry Trevis. This wedding was by licence, in Ashton-under-Lyne. Trevis, described on the licence as being of the 'parish of Manchester', was a member of a Lancashire family with lead-mining interests in Derbyshire. At the time that he was barmaster at Brassington, the office was a gentlemanly one, and lucrative. The seventeenth-century yeoman and gentry families intermarried: Robert Westerne's wife Anne was the sister of the German Buxton who died in 1652, and another of German's sisters was Elizabeth Knowles. Thomas Westerne's daughter Mary married successively James Hall, William Greatrax and William Blackwall, all gentlemen. Blackwall, an innkeeper like his father-in-law, succeeded Trevis as barmaster. Greatrax was described in different records as being of Buxton, Hopton and Brassington, and had property in all three places. Robert Westerne was the only one of Thomas Westerne's sons to outlive his parents, and inherited most of his father's property in Brassington and Carsington. He sold the Carsington property for £145 to Sir John Gell in 1647[13], and the family name ended in Brassington with his death in 1678 and his son Robert's in 1679. The only probate documents for the two men are letters of administration, and without either wills or inventories it is impossible to be sure about their circumstances at the time of their deaths. However, there is the evidence of the ship-money assessment of 1677 that some calamity may have overtaken the family. Robert had been assessed for four hearths in 1772, but his descendants were 'equally assessed' for only 7d in 1677, among the lowest tax rates in the village.

LIFESTYLES OF THE WELL-TO-DO

The Lanes, Buxtons and Westernes were among a group of yeoman/gentleman families whose style can be seen because members of most generations left wills. Others were the Allsops (eight wills between 1535 and 1700), the Charltons (five), the Grattons (six), the Goodwins (three), the Knowleses (six) and the Waltons (three). Only two wills by members of the Wilcocke family were proved at Lichfield before 1700 but this family too was among the leading group. The two wills left by John (1594) and George (1637) make this plain, and George Wilcocke's will named three Buxtons among his 'kinsmen'. One of the longest-established families died out in the male line with the death of Thomas Tissington in 1668.

At the beginning of the century the gentlemen and yeomen lived in houses with one living room, the hall or 'house'. This contained tables and chairs, utensils and fire irons, and at that time was the only room in the house with a fireplace. Richard Walton's 1629 inventory in fact calls it the 'fire house'. Thomas Westerne's hall had a table and two chairs, three 'joined stools', properly jointed that is, and articles to be proud of, one form and 'two boards for stooles', together with nine cushions for his family and customers. The very full list of fire irons – three spits, two brandrethes, two landirons, two pairs of pothooks, tongs, potracks and gallows – indicate the great hall fireplace which heated the whole building and cooked the meals. There was no kitchen. The pots were hung on the brandrethes or gallows over the fire burning in the landiron. Also in the hall were the salts (salt cellars), a kimnell (wooden tub), a kneading trough, barrells, loomes (tubs or vats), kitts (wooden vessels for milk, butter and other food) and a 'hand millne to grind mealte (malt)', which Anne, her daughters and their servants needed for brewing, baking and storage. The hall also contained a store of sheets, table cloths, two dozen napkins and six pillow 'boards' (cases).

The smaller houses of the yeomen farmers became comfortably furnished in the seventeenth century. George Wilcocke's equivalent to Westerne's hall, his 'house', had two tables, two buffet forms (jointed forms for use at a long table), three chairs, three stools, a cupboard and a 'dishboard'. Among the well-to-do, the table in the house or hall was

One of the grander seventeenth-century houses, Sundial Cottage at West End.

big enough to seat a large family. Richard Goodwin's two tables required table cloths 'betwixt three and four yards loung', and 'betwixt four and five yards loung' respectively, bequeathed by his widow Joane in 1686. Wilcocke had three flitches of bacon and three of beef in his 'house' at the time of his death and this impression of an all-purpose living room is given by other wills, though only William Kempe's house in 1641 housed livestock – poultry. The only evidence of decoration in these rooms is in the inventory of German Buxton's goods in 1616, which has 'certaine old painted clothes'.

The parlour was the bedroom of a single-storey house. First-floor bedrooms of two-storey houses were generally known as chambers. The blacksmith Henry Spencer's house contained a 'house', parlour and chamber in 1632. The use of each room is clear from the inventory – 'one bedstead standinge in the parlor & one square table standinge in the dwelling house' and bedding in parlour and chamber. This house's evolution during the seventeenth century can be traced in the manor-court book. Henry Spencer's son Thomas, blacksmith at Tissington, first let and then sold it to his son Anthony. Anthony left it to his son William, who sold the house to Richard Gratton in 1659, immediately on Anthony's death. The entry in the court book recording this last transaction has an agreement that a door 'shall be sett at the north side of the house', Henry Spencer providing the stone and Richard Gratton building it. The mention of stone probably means that this house at least was built from the same material, local limestone, as the houses now standing in Brassington. When Gratton, who was a lead miner, died in 1670, his inventory listed goods in the house and parlour and in the chambers over each of these ground-floor rooms.

Building, extending and improving went on through the century. The evidence of the inventories is strengthened by the Hearth Tax returns. German Buxton's 1686 inventory names parlour, kitchen, house, chambers over kitchen and house and a third chamber which contained nine bedsteads and two trundle beds, which were low and could be pushed under the normal high beds when they were not being used. This chamber may have been sleeping quarters for Buxton's servants – he was a considerable farmer, with oats, barley, hay, cattle, sheep and pigs listed in his inventory, together with equipment which included eight ploughs. Buxton's was clearly a large house and the tax return adds the information that four of the rooms had fireplaces. This was a far cry from the little places in which Buxton's sixteenth-century ancestors had lived.

The same picture of a large well-equipped house was given in Anne Lane's inventory in 1674. The rooms were the house, including a space behind the 'chimney back' for 'two loomes and one great kan', a parlour, a kitchen, a buttery, a bakehouse and three first-floor chambers. Anne's inventory is detailed and revealing – no household list could be more comprehensive than one which included a clothes line, or 'one line to dry clothes upon'. It gives a fair picture of the style of the comfortable yeomen of late seventeenth-century Brassington. The beds included two panelled ones and the inventory lists all the bedding for each. As an example, one of the two in the chamber over the parlour had a chaff bed (mattress) and a feather bed, two pillows and a bolster, three blankets, two coverlets, 'one suite of yellow hangings' and 'five bedstaves', perhaps to support the hangings. There was also one 'mat'. This lay on the rope net formed of cords threaded tightly through holes in the sides and ends of the bed frame. Also in this room was a coffer, and its contents were listed – 'one cofer one red cradle cloath two greene cushion toppes one bed hilling [covering] one red cradle rugg [covering] one old greene carpet [for covering furniture, not floors] one new blanket one pin cushion one hand towel and eight napkins all in ye same cofer.' Cushion covers, towels, table cloths, pillow cases, napkins, coloured cloths for their children's cradles, all demonstrate some style and comfort in the Lane household.

Bedrooms were still store rooms and in the chamber over the Lanes' parlour the appraisers listed 'a box some old bookes in it' and forty-one cheeses, ten bushels of barley, beans, malt and 'blencorne', or wheat and rye sown and harvested together. In an adjoining chamber, among the beds and bedclothes, were 'one old side saddle & one old padde', baskets, hampers, a cradle, sieves, the clothes line and 'one peice of course linen cloath for miners'. The Lanes seem to have been in business for more than just farm goods, as the forty-one cheeses suggest. The cloth for miners' (and others') working clothes was also, presumably, stock-in-trade. The Lanes made woollen cloth: among the articles in the chamber over the kitchen '& within ye lathes' (stored in the roof), were spinning wheels. Also in this room were a vat and a pitch pan, for boiling the pitch used to mark sheep.

The main room, the house, was also used as a store room. In addition to a table, two forms, three chairs and four stools, it contained three flitches of bacon, no doubt hung from the beams, and five stones of wool. There was one luxury item – a 'looking glass'. The house had a fire and a chimney, but the cooking was done in what seems from its contents to have been a large kitchen. It had four shelves, three dressers, one dish board, two forms and a cheese press. All the apparatus for making cheese and butter was there. There were 'costills', large bottles with ears which were carried to the fields suspended from a man's belt, and the Lanes' retail business was no doubt why they had 'one payre of butter waightes' and 'one seaven ston waight one half ston waight two pound stones'. There was a bakehouse with a table and 'one new table leafe and two tressles in ye barne to it', though the baking equipment was in the kitchen – not an oven, but two 'bakestones' and a 'bake sprittle'. The bakestones were flat stones or iron plates on which oatcakes or bread were baked over the fire. The bakehouse had a good supply of cooking fat – 'twenty-seaven pounds of swine grease.'

Brass pots and pans, including frying pans, are common in the inventories, as are pewter dishes. There are, however, few mentions of cups: a pewter cup in Robert Gratton's inventory in 1647 and another in Henry Trevis' three years later. Trevis' inventory also includes a 'pynt pott', but the general rarity of metal or pottery eating or drinkingware suggests that most of the eating and drinking in Brassington up to the middle of the century was from wooden vessels. Anne Lane had three wooden plates as late as 1674.

The twelve trenchers listed in Richard Walton's inventory, along with eight dishes and four 'boles', all valued at 2s 6d, were wooden and must have been what Walton, his family and their customers usually ate and drank from – wooden bowls were for drinking, trenchers were wooden plates. Walton also had two pewter cups. He was a prosperous farmer and innkeeper and his house was well equipped for his time. The cups were perhaps only to be found in houses such as his at the time of his death in 1629.

The houses were lit by candles, and there were brass and pewter candlesticks listed in the villagers' inventories. Anything of silver was found only in the houses of the well-to-do. In an age when outdoor sanitation was usual, Trevis' rooms had items unique in Brassington – 'two pewter chamber-potts 2s 4d'. Trevis had another item rarely mentioned at the time, a warming pan, perhaps the same one as had been listed in his father-in-law's inventory.

'Siled' or 'seiled' (panelled) furniture became common among the better-off in the seventeenth century. Ralph Walton left a 'siled bedstead' to his son Richard in 1636. Beds were the furniture most often embellished by panelling; other items included chests and chairs. Robert Gratton had a panelled chest and chair in his bedroom in 1647. A further refinement to the beds is revealed in the inventories, like Anne Lane's, which mention bed curtains and vallences.

When old Thomas Tissington died in 1668, the last of his line, he gave an unusually detailed description of his farmstead, left to his daughter Alice Travis (Trevis). Tissington's careful definition of Alice's inheritance tells us what the exterior of seventeenth-century yeomen's homes were like. The house was at the west of the village, facing south and near to an enclosed field called Banwoods. The outbuildings attached to it included a bakehouse, a stable and a pigsty and there was also a 48ft-long barn. On the west of the farmyard, Tissington had an orchard of 'longe-peare' and 'round pear' trees, and at the north of the house a garden. The buildings, yard, garden and orchard were all at the side of the village street as farms were until the nineteenth-century enclosures, when new ones were built on the outlying allotments of common land.

Very few of these prosperous villagers included reading among their pleasures. Henry Trevis left 'my Bible and a book which is part of Perkins worke', perhaps showing himself a serious-minded Puritan. William Perkins (1558-1602) was a popular and influential Calvinist writer[14], who preached equality and salvation by good works. Perhaps Trevis' book was Perkins's most famous one, the *Armilla Aurea*, which was published in 1590 and achieved fifteen editions in twenty years. Trevis' brother-in-law, William Westerne, had an entry in his inventory in 1635 which implied that he was an exception in the village in having a small library of books – 'all his bookes and a prospective glasse.' A prospective (or perspective) glass could be either a telescope or magnifying glass or a fortune teller's crystal ball. Westerne seems more likely to have used the former than the latter. Apart from Trevis and Westerne, the only villager whose house contained a book in the first half of the century, on the evidence of the inventories, was William Kempe, who died in 1641. Kempe had a Bible. These were more common later in the century.

Eleven of the villagers who died between 1651 and 1700 had books in their houses. All were from yeoman families with the exception of the minister, George Beresford. The minister's books, valued at 25s, presumably included his Bible. This valuation was much higher than the books in any other inventory and it can be assumed that Beresford's library was for use. In some cases there seems to be cause for doubt – both Edward Lane and his widow Anne were unable to sign their names and their illiteracy was the likely reason for the way in which they stored their books – 'one box some old books in it.' It was perhaps the rise in the number of books being published by the late seventeenth century that prompted these village notables to add books to their furniture, but they were a very small minority, and in at least two of the houses the few books were not read.

PUTTING THE LAND TO GRASS

By the time that Thomas Tissington was making his will there had been a movement among the Brassington farmers from sheep and arable farming to cattle raising; this is according to a document prepared in 1667 in connection with a dispute over non-payment of tithes, in which, incidentally, Tissington's son-in-law Henry Travis was a leading rebel.[15] While most of the inventories from the second half of the seventeenth century include five or fewer cattle, there are six listing twenty or more. The tithe document states that some villagers had converted arable land to meadow, which had the double attraction for them of incurring lower tithe payments than arable and needing less labour, the latter important because 'the poorer sort who are numerous imploy theyre labour in lead grounds.' It was also logical to abandon arable farming on soils as thin as Brassington's. The improvements in arable farming taking place elsewhere would not have improved the farmers' incomes as much as switching from arable altogether. There is evidence in the inventories for Anne Lane (1674), German Buxton (1686) and John Tomlinson (1692): Anne Lane's 'ten strikes of barley beans a little malt & blencorne' were valued at 25s; Buxton's appraisers put a value of only £13 10s on 'about eighty strikes of oates winoed & in the Chaffe & about six score & tenn thraves of oates in the sheaves.' Sixty thraves of barley were put at £5 (1s 8d each). Meanwhile, a thrave of 'corn' in the inventory of John Tomlinson, a miner, was priced even lower, at 1s. These valuations may be underestimates, but it is known that appraisers were more consistent with farm animals and crops than with household goods. It is probable that, in spite of the obvious undervaluing of the household goods of, for instance, Anne Lane, the valuations of the cereal crops are near enough to indicate a good reason for converting ploughland to meadow.

In spite of the testimony of the tithe document, however, Anne Lane's and German Buxton's wills are adequate evidence that there was still plenty of arable farming in late-seventeenth-century Brassington. Its virtual disappearance only came much later, during the nineteenth century, and the oats and barley in such abundance in German Buxton's barns show that there were still fields of corn around the village at the end of the seventeenth century. The implements listed in Buxton's inventory leave no doubt as to the nature of his farming. There were eight ploughs and twenty 'plough-heds' (shares).

SHEEP AND CATTLE

Sheep had formed the main element of the village stock until the latter half of the century. Edward Knowles had 160 in 1628 and his son George, a husbandman, had 140 thirteen years later. There were six others who had fifty or more, and Brassington had plenty of 'wastes and moors' suitable for grazing sheep. The land transfers in the manor-court books refer characteristically to a man's beastgates in Over and Sides Pasture, and to his sheepgates on the 'wastes and moores of Brassington'. One mid-century inventory has enough information to estimate a sheep farmer's living. In 1652, German Buxton left a flock of 230 sheep, giving him an income of 2s per fleece. The sheep themselves were appraised at £74 (7s 9d each), and if the valuations are accurate there can have been little profit in sheep farming by then. The flocks diminished as the century proceeded and the export of wool to the Continent fell. The tithe document ascribed the fall in the number of sheep to the farmers 'letting their meadows for winter pasture for miners' cows', and said that of the village land '35 oxgangs have not one sheep belonging unto them.'

The seventeenth century saw the disappearance of one animal which had been vital to the village's existence since its foundation: there were oxen working the fields at Brassington until the second half of the century, when they were finally ousted by horses. The fall in their use can be seen in the fact that of the fifteen people whose inventories include oxen between 1535 and 1650, thirteen died before 1630. The total number of oxen in this period was forty-three. Thirty-two people in those years owned sixty-three horses, and mention of a 'hackney [riding] sadle' in William Westerne's inventory and of '2 pacsadles' in William Adam's will are reminders that horses were more versatile than oxen. It was only the bigger farmers who owned either horses or oxen during the sixteenth century and earlier, and this remained true of oxen in the seventeenth. During this century, however, some of the smaller men owned horses and it seems likely that by then the oxen were used only on large arable holdings, where their slowness was counter-balanced by their superiority in the heavy work of plough-ing, while horses, cheaper to buy and needing less pasture, were used for riding, light transport and harrowing. The survival of oxen into the seventeenth century may also have owed something to their value for meat. Whatever the reasons, these very large and very powerful animals, yoked in teams of two, four or six, depending on the ground they were ploughing or the weight they were hauling, continued until the second half of the century the work they had done since Brand's men had relied on them to break the ground in their newly-cleared fields.

THE SQUIRE'S HOUSE

By the middle of the seventeenth century, the upper layer of village society, the yeomen with their flocks and herds, fields of corn and their comfortable, well-stocked houses, had the village squire living among them. The will of John Buxton, who died in 1690 'of great age', and the inventory of his goods, drawn up in 1703, show the gulf between the yeomen and the gentlemen. Buxton left property in Ashbourne, Buxton, Compton, Hognaston and Brassington, including his part of the Brassington Manor, the former duchy-manor. He lived in a house, probably the one now known as Brassington Hall, boasting seventeen rooms and two cellars. The rooms were generously furnished with high-quality furniture including, for instance, in the hall chamber, 'a Black Japan Glas, Table & Stand' (varnished, and fashionable – the varnish originally came from Japan), and 'A Stand, Glasse & Sett of Dressing Boxes'. The cellars were well stocked with hogs-heads and barrels, the brew house had a full set of brewing vessels, and the kitchen had plenty of pots, pans, dishes and cooking equipment. The amount of brewing equipment, the eighteen barrels and hogsheads, and the 'twenty dozen of glass bottles' may mean that Buxton, like Thomas Westerne and the barmaster William Blackwall among other earlier gentlemen in the village, was an innkeeper.

To end this family's story in Brassington, John Buxton's son Richard, not yet born at the time of John's death in 1699, died in 1724 after a short and what seems to have been an unusually extravagant life. He left all the property which had come from his father in trust to his friends John Addinsall Leake of Wirksworth, gentleman, and John Mellor of Ashbourne, innkeeper, instructing them to pay his debts. Anything left after these trans-actions was to go to Richard's cousin William Newton. Among the property which Newton in fact received was the Buxton part of the manor of Brassington.

THE WORKING VILLAGERS

The squires and yeomen shared the village with the people whom the tithe document of 1667 called 'the poorer sorte'. The Brassington parish register of baptisms, marriages and burials begins in 1716. Before that date there are the entries preserved, with gaps, in the bishop's transcripts – parchment copies of the register entries sent by the vicar or his curate to the bishop in Lichfield. Before 1673, even these 'short and simple annals' of the landless villagers have been lost. A few of them made wills and some appear in their masters' wills. Their names are listed in the Hearth Tax returns, especially the one

Brassington Hall, probably John Buxton's seventeenth-century home.

for 1664 which gave the names of those too poor to be liable for tax. They appear in the manor-court records, occupying, that is renting, houses and fields, and paying fines for fighting, letting animals stray or getting into debt. Even in the days when every man had a portion of ground and a share of the manor pasture, there were those whose holding was too small for a living and who became wage labourers, doing the work on the lord's demesne land formerly done by the copyholders. Some of these landless men became craftsmen, vital to the village. Some of the craftsmen prospered enough to buy themselves back into the group with small land-holdings. Henry Spencer the blacksmith was an example: he bought a copyhold cottage from Edward Knowles in 1614 and a further cottage and garden and 'a little yard' from Roland Allsop in 1620. William Westerne was a witness of his will in 1633 and his son Henry was left part of the Westerne property in 1635. However, most of the labourers lived precariously, dependent on arbitrary economic conditions.

Queen Elizabeth's successors tried hard to enforce the Poor Laws and required local justices of the peace to make returns to the Privy Council in London of what they were doing to administer them. In 1631 the justices of the Wirksworth wapentake reported:

> We have taken care that the Lords and parishioners of every town relieve the poor thereof and they are not suffered to straggle and beg up and down either in their parishes or other where. But such... as have transgressed have been punished according to law...We have one house of correction at Ashbourne, within our wapentake, where such as are committed are kept to work...We have punished several persons for harbouring rogues in their barns and outhouses.[16]

The harassed parish overseers took pains to ensure that their responsibilities did not extend to visitors from other parishes, who were sent on their way. This system was strengthened by the Act of Settlement of 1662. Without a certificate of settlement, no one was allowed to stay in the village, and the manor-court played its part in getting rid of unauthorised 'incomers'. In 1666, 'Item we lay a payne that yf John Lane Jnr doe not remove Martin Hallom his inmate before Michaelmas nexte to forfeit xxs... Item we lay payne that if John Rippon doe not remove Candish Heneley his inmate before Michaelmas nexte to forfeit xxs.' Lane and Rippon had already been fined 3s 4d each for lodging their 'wandringe rogues' and William Rippon was fined 6d for having 'vagabond psonas in *domo suo*' (in his house).

One of the more constructive duties of the justices was to put poor children to work with farmers or tradesmen and to make their master responsible for their welfare – 'We have made speciall enquiry of such poor children as are fit to be bound apprentice to husbandry or otherwise and of such as are fit to take apprentices & therein we have taken such course as by law is required. And we find none refuse to take apprentices being there unto required.' The justices made attempts to prevent the apprentice-masters exploiting this cheap labour. At the Quarter Sessions in 1703[17] 'it is ordered by this Court that Richard Gratton of Brassington take care and provide for Stephen Barton his Apprentice lately put to him by Indenture as hee will answer the contrary at his peril.' Stephen Barton's sister Ellen had also been indentured to Richard Bennett, a Brassington husbandman. Bennett successfully appealed, and: 'it is ordered by this Court That the said Indenture be vacated and discharged.' The grounds for Bennett's appeal are not given, but he was a bachelor in a small way of farming and may have convinced the court that he could neither train nor adequately maintain the young girl. One of the witnesses of Bennett's will when he died in 1715 was Samuel Barton – Ellen may have been a relative of Bennett and Bennett considered a suitable master for that reason. Whatever the background, the case of Stephen and Ellen Barton shows that the apprenticeship system was not always successful in dealing with the problems of a poor family.

The villagers could appeal for help directly to the magistrates. George Woodiwiss, miner, did so in 1689.[18] He described how he:

> ...haveing long wrought hard at the lead mines and not at all of a long time gotten any thing from them to quit the charge of them: and now being in great penury and want: and every day expecting to be turned forth of doores for not paying his rent: prays that your good worships could consider the sad condition of your petitioner.

An almost illegible note records that the justices agreed to pay Woodiwiss 6d a week towards his house rent, plus a further 1d a week, but also refers to 'an assault upon Sarah his wife.' Then, as now, poverty put strains on families which were sometimes more than they could bear.

Two petitions illustrate the ruinous effect of desertion by a breadwinner. From the middle of the century comes a petition by a group of Brassington women who included the midwife Jane Stafford, Anne Westerne and Mary Blackwall.[19] They testified that they had been present at the birth of a child to Elizabeth Buclowe, unmarried daughter of Widow Buclowe, and that:

> ...the said Elizabeth did before us women at the labour uppon many oathes and in the extremity of paynes as she bore and brought it forth that the said Rowland Allsop was father to the said child which often promised her marriage but nowe he faileth that promise, and keepeth out of way.

They asked the magistrates to order Rowland or his yeoman father, Robert, to support the two destitute women and the baby. At the end of the century the miner Richard Lee petitioned on behalf of his daughter Jane, deserted by her husband James Guy, 'a very extravigant person', after two years of marriage and the birth of a daughter.[20]

Stephen Barton, father of Stephen and Ellen, had died in 1696. He was one of the miners and other working men who had by the middle of the seventeenth century become prosperous enough to leave wills when they died during the last quarter of the century. The miners Stephen Barton, Thomas Buckley, John Madder, George Steple, John Tomlinson and Richard Walker were all from families whose members had played no part in the government of the village and had never before made wills. These men demonstrated a rise in their status by leaving one, and one of them, George Steple, had in fact served on the manor-court jury, a duty which implied that he held copyhold land. His will describes it as 'one little garden…which I hould under the lords of the manner of Brassington.' Thomas Buckley had grazing right in the pasture (one beast-gate) and must therefore also have been a copyholder; Stephen Barton had a close called Catgreave; John Tomlinson, who left a pig and two cows, plus hay and corn, and Richard Walker, with two 'melch cows', presumably rented their pasture, since they left none of their own. These men represented a group of independent miners/farmers characteristic of Brassington. The poor husbandmen Robert Wayne, Anthony Coates and John Mellor may all have supplemented their farming income from lead mining. It is certain that many more than the few who left wills lived on a mixture of lead mining and farming, keeping a few cows and pigs on rented land, using the wastes for pasture. The 1667 tithe document is explicit: '…the poorer sort who are numerous imploy theyre labour in lead grounds in ye Lopps [lordships] 4 or 5 miles round them, who will give great rates for meadow grounds to keep theyre cows in winter.' They were not paupers, but they were poor, and the fate of Stephen Barton's children after his death, and of George Woodiwiss after a period of unsuccessful mining, shows how precarious their livelihood was.

Some of the poorer people seem from the inventories to have lived in houses with the 'house', parlour and two chambers arrangement. John Madder, for instance, left his daughter two rooms in his house while she remained unmarried – a bedroom and a second room for her possessions. This sounds like two up and two down. In another case, Robert Wayne left 'fowre bedds with the furniture', strongly suggesting two bedrooms at least. Most of the houses were smaller than these. George Steple, the miner, seems to have had a two- or three-roomed house. He was assessed for a single hearth in 1662 and omitted from the tax list altogether in the later assessments. His inventory in 1685 confirmed the fireplace by listing his fire irons. However, there is no mention of rooms, and it is probably significant that he had only one bed in his house. Another miner, William Scattergood, was listed among the 'persons not chargeable' in the 1664 Hearth Tax return. When he died in 1700, his inventory was explicit about the nature of his house. He left goods valued at £4 in three rooms – house, parlour and chamber. Many of the surviving houses in Brassington are from Scattergood's time. Most have been extended, amalgamated or otherwise adapted to suit changing circumstances and fashions. However, there is at least one which looks as though it has survived unchanged: this is a small house near to the former George & Dragon pub. It has mullioned windows, carved out of limestone, and is very low. It has been uninhabited for many years and used as a store shed. This limestone cottage is tiny to modern eyes, but in the seventeenth century comparison would have been with its one-roomed, clay-walled predecessors.

These poor houses were better furnished than most of the houses had been in the previous century, sparse though their household goods are when compared with such as the Lanes'. Madder's little house contained brass cooking pots and pewter dishes, a cupboard and 'three little coafers', three chaff beds with pillows, sheets, blankets and coverings, tablecloths and napkins, shelves and a dish board, chairs, stools, a form, a frying pan and fire irons. Some inventories are uninformative about the separate items of 'pewter and brasse', 'irons about the fire', 'chaires and stooles', but Madder's house was furnished in a manner typical of the small husbandmen and miners who left wills in the latter years of the seventeenth century. There was one house, however, whose furnishings

Mullioned window of a former labourer's cottage, Miners' Hill: one of the poorer houses, with rough stone work and mullioned windows on a stone shed, it was probably a dwelling for most of its life.

may exemplify the style in which most of the villagers lived (the large majority who left no will). This was the three-roomed house left by William Scattergood to his wife Katherine in 1700. The 'house' contained a table, a dish board, a cupboard, one seat (presumably a form), and 'puter and Brass' – a total value of £2. The parlour had a bed, two coffers and a box at 20s, and in the chamber were two beds, two coffers and a kimnell, also put at 20s. There is no mention of tablecloths, napkins or other linen. The appraisers seem to have found no other bedclothes than those on the three beds, and the Scattergoods had even less cooking and storage equipment than the Madders. This sparsely-furnished little cottage must have been the usual dwelling of most of the people of Brassington at the end of the seventeenth century: however, no doubt there were some who lived in worse circumstances. Sharing the poverty of the labourers and the unemployed was at least one poor scholar, George Beresford, described as 'late Minister of Brassington' in his inventory in 1693. Beresford's sad little list of possessions included one bed at 9s, 'other small things' at 7s 6d and 'other small things out of sight' at 2s 6d. However, his books were valued at £1 5s in an inventory whose total value was £2 17s and we can perhaps assume that he, like his appraisers, valued them the highest of his possessions.

Many of the small cottages have survived. The evolution of one of them can be seen in its gable end, just over the eastern wall of the churchyard. An originally three-sectioned mullioned window has been given a central bar to convert it to a two-sectioned arrangement and the right hand of the two sections has been filled in, probably by an eighteenth-century villager avoiding Window Tax. This cottage was built in the late seventeenth or early eighteenth century and adapted and modified many times. It was extended in the late nineteenth century and altered again recently but still looks like a Derbyshire lead miner's cottage, as do most of the houses in the village.

Former miner's cottage, Hillside.

Many of the families which can be traced in the parish registers through the eighteenth and nineteenth centuries had been in the village in the seventeenth and sixteenth. Those who held land were, as we have seen, well documented in the records of the earlier centuries. Those who had no possessions are occasionally seen too. The John Banks who was fined for building a cottage on the waste in 1652 was a member of a family who lived in Brassington for several centuries and who were substantial farmers in the eighteenth and nineteenth. The Taylors had been in Brassington at least as early as 1410, when the name appears on the duchy-manor rental. They were landless in the sixteenth and seventeenth centuries, when the name surfaces occasionally in the court records. They were still in the village three centuries later. A Thomas Slack who made his only appearance in a record as an appraiser of his brother-in-law Richard Jackson's inventory in 1637, is a possible ancestor of two men, father and son, with the same name, who were barmasters in the eighteenth and nineteenth centuries. The best evidence of the long-surviving poorer families in the village – those who paid no tax, owned no property and left no wills – is the presence of their names on the manor's suite rolls. The one for 18 April 1664 has, in addition to the ones already mentioned, Thacker, Wright, Bacon, Rippon, Kirk, Roose, Eaton and Lee, all names which can be found in the parish registers, mining records, manor-court books, wills and other records in the next two centuries. Family names died and families moved. But many of the families at the turn of the nineteenth century had been there for several hundred of years, evidence for their existence in the record of a member being fined in the manor-court, transferring an acre or two of ground or taking his place with his neighbours in the list of those 'oweinge suite and service' to the lord of the manor.

REFERENCES

1 NA E179/245/7

2 NA E179/94/46219/94/402

3 Edwards, 1982(1)

4 NA E178/245/9 (Roll 3)

5 Edwards, 1982(2)

6 DRO D161 B/6/52

7 MLSL BM Add MS 6690 f3; DRO 166M/M1

8 DRO D258 Box 29/13

9 MLSL BM Add MS 6690 f3

10 DRO D258 54/1sb

11 Dugdale, 1989

12 DRO D258/54/14j

13 DRO D258 44/8c

14 DNB

15 DLSL Gale Bequest bundle 3

16 VCH, 1907, p178

17 DRO QS Book 2

18 DRO QSB 404

19 DRO QSB 137

20 DRO QS IX crime box petitions: III,141

CHAPTER 6

MANOR, CHURCH AND STATE

SQUIRES AND TENANTS

Ten years before William Savile's purchase of the Shrewsbury Manor, the king, Charles I, had given away the duchy-manor to a certain Charles Harbord[1], a move which probably reflected the Duchy of Lancaster's estimate of its worth, tenanted as it was by villagers paying very low and fixed rents for their copyholds. Harbord sold the manor in 1632 to Edward Pegge of Ashbourne and some of his associates.[2] Pegge had been the manor's steward for over ten years and its attraction for him may have been the prestige of being lord of the manor rather than any financial gain. He soon parted with a quarter of it to John Buxton and its fragmentation continued when Buxton's son Richard sold this quarter to Savile in 1649. The price was low – £150. A later transaction saw Savile and George and Henry Buxton buying a 'moiety' of the reduced manor in 1652 from Pegge and two associates.[3]

By 1640 then, the people of Brassington had local gentry for landlords. Most of them were absentees, but members of the Buxton family, gentlemen and lords of the old duchy-manor until Richard Buxton's death in 1724, lived in the village and could fairly be described as the village's squires. John Buxton's probate documents show that they lived in considerable style, and they had their coat of arms displayed on their pew in the church – 'sable, two bars arg.; on a canton of the second, a buck, trippant, of the field. Crest: a pelican, vulning itself, or.' What did the two manors amount to in the seventeenth century, and who were their tenants? As we have seen, the manors were not geographically distinct. Their meadows, lands and, later, their enclosed fields, were scattered and intermingled. They were of roughly equal extent, but differed markedly in the way the tenants held their land and in the rents they paid for it. There are rentals for the duchy and Pegge's manor for 1620 and 1640 and for the Shrewsbury and Savile's manor for 1619 and 1639, making comparison easy.

Very clearly the duchy-manor kept the old ways and the old tenant families. For instance, the four oxgangs of copyhold land which George Wilcocke held in 1620 had by 1640 passed to his son William, who was paying the same rent as his father had paid twenty years earlier. In 1640, Thomas Tissington, the last of a very long line, still had the same holding as his ancestor John had had in 1560 – 3 bovats (oxgangs) less 3 acres. He was certainly paying the same rent for it as his father, another John, had paid in 1620, and probably the same as his Elizabethan ancestor. The names on the rentals of 1620 and 1640 are the same – Wilcocke, Gratton, Knowles, Allsop, Buxton, Toplis and Tissington. The only change is William Greatrax for his dead father-in-law, Thomas Westerne.

Westerne's large holding in the neighbouring Shrewsbury Manor in 1619 was still in the family in 1639, held by his son Robert. Apart from the Westerne family's continued presence on the rental, however, and a chief rent for freehold land in the manor which had remained at 26s, the rentals for 1619 and 1639 look quite different. There had been changes in the tenancies from one family to another, and it can be assumed, though the rents are not given in the 1639 rental, that the new owner, Savile, was collecting very much more rent for his land than the £5 per oxgang of 1619, which was itself twenty times the duchy-manor rent at the time.

A duchy rental for 1620, drawn up and signed by Richard Buxton, 'the Reeve of Brassington and accounted for by him at the audit at Tutbury', is annotated 'twenty-eight oxgangs of copie hould land in Brassington.' This rental, slightly different from the one accompanying the custumal of the same year, has separate lists for twelve copyholders, for six 'cotagers that have noe oxgangs of their owne', for four freeholders, and for six 'outrents', which include 'and thirteene hens' payable by Elton people, in addition to a rent of 6s 8d. The copyholders' rents come to £7 14 10d and the rest to 52s, a total of £10 6s 10d. By 1640 the manor, by then owned by Pegge, still had twenty-two village tenants, plus the neighbouring villages' payments for their grazing rights, plus Sir John Gell and Sir John Curzon's payments of 6d and 6s 8d for Longston Fields and Ivenbrook Grange respectively. The total rent was £9 4s for twenty-eight oxgangs, over thirteen cottages and one house. The list of copyholds included seven named enclosed fields – Caldwell Sitch, Yards, Ragstaff, Crosse Green, Elme Yard, Pater Noster and Little Green, all in the area to the south and south-west of the village where the arable strips had been.

The farmland in Brassington, in 1640 the twenty-eight oxgangs of Pegge's manor and a similar amount, plus two freeholds, in the 1639 rental of Savile's, amounted to about 1,000 acres. This was held in the middle years of the century by thirty-seven tenants and freeholders. Some of this land, and many of the houses held in the village by manor tenants, were rented from them by landless villagers. The manor land included the village itself and the arable, meadow and pasture near it, and amounted to about a quarter of the total acreage within the parish boundary. The rest was common land – the 'moors and wastes'. When the Duchy of Lancaster confirmed the Brassington tenants' copyholds in 1620, it listed the 'several beast pastures, commons and wastes' where they could graze their animals.[4] They were 'the Great Beast Pasture and Sydes' on the eastern slope of the valley in which the village lies, and 'a piece of ground in Brassington, commonly called the Green, in which there is a well or spring of water called the Coole Well.' This was at the southern end of the village, by the highway. The wastes were in the north. They included 'the hill or parcel of hilly ground about Brassington church-yard whereupon stands a rock or tor of stones, commonly called Earnest stones.' This was also called Ernstone, or Yernstone, and the lane running past it into the Pikehall Road was formerly called Yernstone Road or, as older Brassington villagers called it, Genstones. The list of wastes has Manystones, Roundlowe, Longcliffe, Street Knowl, Slipperlowe, Gallowlowe and more, all in the north.

This moorland was not to be enclosed until the early years of the nineteenth century, but it seems likely that by the time the manors were sold to the new lords most of the ancient arable land had already been enclosed. The enclosure award of the nineteenth century[5] refers to the fields in the south and south-west as 'ancient inclosed lands', and a glance at the map or, better, at the fields themselves, makes it plain that these fields were formed from the original strips. The field boundaries, here consisting of hedgrows, are quite unlike the geometrical stone boundaries of the later enclosure. Their curves follow the lines of the old strips, and the strips themselves are easily to be seen as linear hollows and ridges. Many of these fields had been made in the early years of the seventeenth century by villagers exchanging their strips to consolidate their holdings and incurring the displeasure of the Duchy of Lancaster for doing so without the permission of the manor-court. Enclosure of the demesne land of the duchy-manor had by a meeting of the court in 1641 proceeded to the extent that the jury had difficulty in saying where it was. They claimed that they 'have heard of' demesne land in a number of closes. For instance: 'wee present and have heard that there is pte of the Kings Demesne lyeing in the Close called Washehills in the occupacon of Robert Westerne.' Clearly, by the time that Edward Pegge and his two associates, John and George Buxton, set about tracing where their newly-acquired demesne property lay, most of it had been enclosed by their tenants.

In 1667, according to the tithe dispute papers, 'all ye fields of Brassington have been inclosed about 50 yeares, except part of a field wch was inclosed 30 yeares since or therabouts.' These enclosures were not carried out unopposed by the villagers, who saw their rights of common pasture disappearing. The same papers include a statement that in or about 1653 William Savile had enclosed a 'sheepwalk', or in other words had converted part of the arable land to a fenced sheep pasture, 'not wthout a tryall at law wth ye townesmen who opposed him.' Enclosure was in fact incomplete by 1670, and a manor-court penalty in that year against leaving mine shafts open referred to, 'ye lands of ye said liberty of Brassington either inclosed or uninclosed.' This reference would include the wastes, where most of the mine shafts were, but there were fields in the old cultivated south still not enclosed, and the 1803 Enclosure Act specified 'open fields' in its title. With these exceptions, however, there seems no doubt that by 1670 Brassington was essentially an enclosed village.

THE MANOR COURT

The new lords kept up the old forms, the transactions of their joint court meticulously (and wordily) written up, though by their day the old parchment rolls had given way to books. The Brunt/Taylor arbitration document of 1680[6] reveals where the court held its meetings. It instructed Ralph Brunt to attend the next meeting 'att ye now dwelling house of William Dodds in Brassington where ye Court is kept.' There was no village hall in Brassington and the court held its meetings in one or other of the inns. The jury appointed the reeve and other officers, chief of whom by the seventeenth century was the constable. It continued to present villagers who had broken the rules, but by the mid-seventeenth century a manor-court's power to punish was restricted to fines of less than £2 7s. In the 1660s and '70s, a remarkable number of villagers were awarded fines of 39s 11d. The chief and essential function of the manor-court by the middle of the century was to provide the machinery for the transfer of copyhold land by sale, mortgage or inheritance.

The governments of Queen Elizabeth and later monarchs set up new local government systems, based on the parish vestry, justices of the peace and the Assize courts, and the manor-court gradually lost the role it had had since before the Norman Conquest in setting and enforcing local bylaws. However, the court was still the forum in which the villagers governed themselves. The jury which was unable to say where the lord's demesne lay in 1641 consisted of men who had lived in the village all their lives and whose ancestors had lived there for centuries – Buxton, Wilcocke, Tissington, Charlton, Knowles, Gratton, Allsop. They were equally vague about who the lord's subjects were – 'Wee present to our knowledge wee doe not know whether the freeholders or other of the Tenants in the Towne doe owe any suite or service to the Courte but the Copyholders only.' It is difficult to avoid the conclusion that the jury was exercising passive resistance to Edward Pegge's attempt to be a real lord of the manor.

The 'pains' or lists of dos and don'ts which the courts continued to lay down throw considerable light on conditions in the village. Brassington lies in a north-south cleft in a barren limestone plateau. There is no water on the plateau and the village's wells were vital for the people's wellbeing. The court had the preservation of clean water constantly in mind. Most lists of pains mentioned it, and there were many cases of villagers being fined for misusing wells or taking water from the village. Pain number four in 1661 instructed that 'no psn shall wash clothes, beaste meate or swine meate or any other noysome or filthy thinge att the Coole Well or other wells in the Towne within three or four yards of the wells mouths to corrupt the water.'

Number five forbade miners to take the water away – a prohibition which must have been hard on men mining in a dry landscape, where water was essential to prepare their ore for sale. The fine in each case was 12d, and we can be sure that the villagers continued to use the wells for washing meat and continued to take water from them to the mines. The court's centuries-old anxiety about fencing remained after the open fields had been enclosed, focused now on the common pasture which had been retained. A meeting in October 1671 appointed four men, with the familiar names of George Wilcocke, William Toplis and Thomas and Andrew Lane, to see that 'all the fences belonging to both the over pasture and the Sides pasture shall be well and sufficiently made and walled seaven quarters [of a yard, i.e. 5ft 3in] high.' Any villager not maintaining the part of the fence adjoining his land was liable to a fine of 12d while any backsliding by any of the four overseers was punished by one of 5s. Their inspection was to be done not more than ten days after 25 March, but that the fine was needed to spur the unwilling officers into action is plain from a meeting in April 1672, by which time nothing had been done. The overseers were instructed to carry out their inspection by 5 May.

CRIME AND PUNISHMENT

Crime was dealt with by the justices of the peace, appointed by the government from the ranks of the land-owning gentry. The records of the Quarter Sessions at the Derbyshire Record Office show how the system worked. On 23 January 1648, for instance, William Wheildon, husbandman of Brassington, accused his fellow villagers John Wright and wife Anne of stealing coal from his mine.[8] Note that the husbandman was also a miner. Wheildon stated that he and two 'groovefellowes', Richard Walker and William Tomlinson, had a coe on Hopton Moor. When 'halfe a horseload of coles' disappeared, Wheildon suspected Wright and took the deputy barmaster with him to search Wright's house. They found coal in his parlour which Wheildon claimed to be able to identify because it had traces of 'smitham' or lead dust on it. The Wrights, whose choice of parlour for coal store is revealing, claimed that they had bought the coal from Thomas Johnson of Brassington. Anne used turns of phrase which sound modern – she had 'not been near' Wheildon's coe and had 'never touched' his coal. She also gave the price of Johnson's delivery – 2s a load.

Because mining was involved, Wheildon had called in the barmaster to present his case to the justices. For other offences the constable was the man to help. John Hodgkinson missed five newly clipped fleeces in 1694.[9] He had sheared the sheep in a 'walke touching the houses of severall' and, after a search, the constable found 'a parcill of wooll which was very moyst & seemed new Cutt of the sheeps Back & was unwashed & this wooll was found in Robert Willsons house.' Willson, like John Wright, claimed to have bought the wool and told a convincing story about bargaining for it with the wife of John Palmer before paying a shilling for it. For his part, Palmer 'saith that since candlemas he had killed seaven sheepe' and that the wool in question was from 'a Hog of his that Broke his leg yesterday on Midlepeak.'

Two cases illustrate both the violence common in every period and also the fact that the local gentry were often involved. Both cases feature a German Buxton, though which one is impossible to say. He was the aggressor on one occasion and victim on the other, and clearly spent much of his time in alehouses. In the first case[10], Buxton and two companions left an Ashbourne alehouse one night, walked noisily through the town, threatened a watchman with a 'longe Bill' and insulted the constable, whom they had roused from his bed. The watchman ran away and the constable stayed in his house because 'his wife would not suffer him to go forth, for feare hee should have byn hurt.' Buxton took his friends off to another alehouse 'to get drinke' and promised to return. The Buxton family's recent rise from yeomen to gentry may have unsettled them – Richard Buxton had been fined 3s 4d by the Brassington manor-court in April 1653, three months before German's escapade, for assaulting the village constable, Robert Allsop. German's second court appearance came in 1665.[11] The words of the indictment cannot be improved. Anthony Radcliffe of Glossop, 'a common frequenter of alehouses', was accused:

> that… hee did make two affrayes & bloodshedds upon German Buxton gent without any just cause; And did pull the hayre off from the said Buxton's head, And struck him upon the head… and spurned at the said Buxton his privy members with his foote, haveing on a great heavy paire of boote Cloggs, full of Iron nayles.

ROADS AND TRANSPORT

Squire Buxton, the yeomen and the labourers, once they stepped outside their very different dwellings, found themselves sharing a rough and ready world, rough and ready in more ways than one – the roads, for instance. Brassington's highways in the sixteenth and seventeenth centuries were the present Bradbourne Road, the Wirksworth Road and the road which today runs through the village from Longcliffe to join the Ashbourne to Wirksworth Road. This last was part of the main highway between Derby and Manchester, and Brassington was a regular calling place on it. An Act of Parliament in 1555 had made the upkeep of these ancient highways the responsibility of the parish vestries, though the manor-courts continued to share it: they had been responsible since the Statute of Winchester in 1285, and it was their inability to cope with the growing trade of the sixteenth century that had prompted the new arrangement. The parish was responsible to the justices of the peace at the Quarter Sessions in Derby. It elected overseers who enforced four days of highway work each year from every householder – or were supposed to do. There is evidence that they managed to evade it.

The members of the vestry were probably the same as the manor-court jury. On 27 April 1670, they were German Buxton, John Allsop, Richard Gratton, Henry Travis (or Trevis), William Toplis, William Lane, William Allsop, Daniel Bagshaw, George Jackson, Rowland Allsop, Robert Allsop and Henry Spencer. These were the men with horses and labourers to send to the highway and they would also provide the overseers who drafted in the poor of the village to pick, cart and lay the stones used in road repairs. In 1671 the overseers were Richard Buxton and William Charlton, who were threatened with fines of 5s if they did not ensure that 'every pson to doe their duty in helping to repair.' Repair meant filling in the ruts with stones. The tools used were hammers, shovels and rakes, and there are many contemporary accounts of the very inadequate result. Decent road-making techniques, lost when the Romans left the island, were not to be redeveloped for another century. Hundreds of years of wear had produced the sunken lanes still familiar in many places in Derbyshire and often known as hollowgates or holloways. Brassington still has one, in a now-disused stretch of Middle Lane bordered by fields called Holloway.

The manor-court had to contend with what seems to have been a general disregard for the whole idea of the open road. The court of 27 April 1670 had to deal with Richard Brough and Robert Wright *'quia erexer sterquilinos suos in coi via ad nocumentu'* – Brough and Wright had offended the court, and their neighbours too, we must suppose, by building their dung heap in the road. Along these rutted, rocky tracks, kicking up clouds of dust in summer and sinking into the mud in winter, skirting obstacles such as Brough and Wright's dunghill, went the ox- or horse-drawn wains and the pack-horse trains.

The raw wool from Brassington flocks was probably carried away on pack horses. William Adam, active from the late 1500s, left '2 pacsadles girthes & halters' when he died in 1634. The pack-horse trade has left us the field name Jaggerway to the south of Brassington, a reminder that the road here was travelled by the strings of loaded horses

The former Wirksworth Road at its junction with the main road through the village, Pikehall Road. It winds steeply through the Dale and was closed for through traffic about twenty years ago.

led by their 'jaggers', the men who led the horses carrying their jags, or loads. Lead ore seems to have been carried in waggons where it was possible to do so and Derby and Chesterfield men were carting ore for export from Southampton as early as the mid-fifteenth century.[12] No doubt the ore had reached the Chesterfield and Derby buyers from the ore-fields, including Brassington, by pack horse. A normal pack-load in Queen Elizabeth's time was about 2cwt while two horses could draw a cartload of a ton, making carting more profitable where it could be done. That it was not always possible is shown by a remark in the 1667 tithe document that corn from Brassington had to be carried on horseback, 'there being no passage for carriages amongst the torres and rocks.' During the sixteenth century, when the Earl of Shrewsbury was exporting lead to the Continent from Bawtry, Brassington lead went there on pack horses.

TRADE

It is apparent from the sixteenth- and seventeenth-century inventories that most crops were grown and animals reared in Brassington for subsistence. However, there would always have been small numbers of cattle and larger flocks of sheep driven to and from the market at Wirksworth, sharing the road with wains, pack-trains, horsemen and walkers. While the men with two or three cattle would buy from the Buxtons, Lanes, Tissingtons or one of the other larger farmers, they would no doubt have occasion to drive a cow or a few sheep to market. The farmers who moved from arable to cattle farming at the end of the seventeenth century would buy and sell regularly as part of their business and their stock would often be among the herds on the drives to market. There were a few farmers whose crops seem large enough to have provided surpluses for sale, though it is difficult to estimate from the inventories how great these surpluses of wheat, barley, oats, beans and peas were. Most was certainly for home consumption. Ale was brewed and bread baked in the village. According to the rental of 1620, there were no mills in Brassington and the wheat must have been taken elsewhere for milling, but the villagers baked their own bread or bought it in the village, and it is difficult to see much surplus wheat for sale. Some of the crop was perhaps sold to 'swailers' or 'badgers' – travelling dealers in oats or oatmeal who would take part of the oats crop, as well as surplus eggs and cheese. They sold goods from the villages in the Wirksworth market and there were Brassington men in the trade. Samuel Barton and Robert Buckley were licensed swailers in 1725, for instance.[13]

As with brewing, baking, mining and farming, the development of retailing is indicated by the authorities' need to regulate it. The justices of the Wirksworth wapentake included a paragraph in their report of 1631, quoted earlier, which read '[we] have made searches in Market townes & other places & taken away & burnt very many false weights & measures, & taken order for the punishing of the said offenders.' The Kempe family, though described as mercers, and carrying a comprehensive stock of cloth, were obviously running the kind of all-purpose shop which was until recently common in both villages and town neighbourhoods. In addition to their shop – and there were no doubt successors to it when the Kempes went out of business – the inns, bakers and smiths, there were butchers, a tailor, an ironmonger and a carpenter identifiable in the seventeenth-century records.

The ironmonger, Daniel Bagshaw, issued tokens from his shops and Derby Museum has one issued by him in 1669. There was by the seventeenth century a great shortage of small change in the currency. The increase in trade during the Elizabethan period and afterward had increased the amount of money in circulation, but the Royal Mint had refused to issue base metal coins. When it eventually decided to do so, during the reign of Charles I, the project was suspended after his execution, and not carried out until 1672. During this intervening period traders of all kinds issued their own tokens, usually of brass. Daniel Bagshawe's tokens, given by him in change, would circulate in Brassington, and probably in Carsington and the other villages nearby. Traders receiving them would save them until they had enough to exchange them at Bagshawe's shop for silver coins – forty-eight of his brass farthings for one silver shilling. Bagshawe was a member of the manor-court jury and his accomplished signature suggests that he was an educated man, but his business seems to have been a modest one, judging from his inventory, where his stock was valued at £10.

CIVIL WAR

In the Civil War between king and Parliament, Derbyshire was held by Sir John Gell of Hopton, whose efforts prevented the Duke of Newcastle's Royalist army in the north from joining Prince Rupert to the south of the county.[14] Gell had been created a baronet by the king in 1642 but was a Presbyterian, opposed to the reforms of the Church of England being pushed by the king and in general to the king's absolutist ambitions. He proved a fierce Parliamentarian fighter. The lords of Brassington Manor, Pegge, Savile and Buxton, took the same side.

The king raised his standard at Nottingham and recruited many miners by promising to remove the duties on lead ore.[15] While he was moving through Derbyshire, Gell was in Northampton, where the Parliamentary commander, the Earl of Essex, gave him a commission, with the rank of colonel, to raise a regiment of 1,200 men to defend Derbyshire. After trying unsuccessfully to recruit miners in Wirksworth he went to Hull where he persuaded the governor, Sir John Hotham, to provide him with 120 trained men as a nucleus for his regiment. While Gell was away a Royalist unit moved into Derbyshire, sacked and plundered Gell's home, Hopton Hall, and looted the countryside. A journal of the time, *England's Memorable Accidents*, in its issue covering 31 October–7 November 1642, described how Royalist cavalry 'plundered two houses in Brassington, and threatened to fire at the towne, they gather to them daily Miners and other Shag-ragges.'[16] This experience turned opinion Gell's way and when he returned and drove the Royalists out of the county, he was more successful with his recruiting, while many of the newly-recruited miners left the king's army.

During the years 1642 to 1646 Gell's regiment, which added a cavalry unit to his infantry, was based at Derby and in continual action, moving to all parts of the county and into neighbouring counties to repel threats from Royalist armies in the north and south. The villagers of Brassington became familiar with armed men coming and going and with Gell's efforts to raise money for the war. The accounts of the committee which ran the county on Parliament's behalf show that the six richest farmers at Brassington gave a total of £32. Most of the money received from the county was paid by the committee's treasurer to Gell's commissary (quartermaster), who from January 1643 was Henry Buxton.

THE CHURCH UNDER THE COMMONWEALTH

The side of village life most affected by the victory of Parliament was probably its church-going life, and the fact that Henry Trevis owned a copy of William Perkins's book may mean that the Puritan doctrines had support in the village. The shifts in doctrine and practice which must have been forced on the villagers by successive curates were at their most thorough at this time, and the Anglican Church was reorganised on Presbyterian lines.[17] The parish presbytery, that is the curate, usually called pastor under the new dispensation, and the church elders, had the duty of overseeing the Church's affairs, maintaining discipline among the parishioners and censuring offenders. The presbytery had the power to warn, to suspend or to excommunicate the villagers for behaviour which it disliked, and it is easy to imagine the tyranny which could be exercised by a zealous curate under such a regime – offenders could be forced to make public confession of their sins, and if they failed to mend their ways could be denied the Lord's Supper and prevented from having their children baptised. However, there is a hint that the curate at Brassington during the Commonwealth may have been independent-minded enough to stand against the trend. A Parliamentary Commission of 1650 recorded that 'Mr Thomas Alsop is curate and scandalous.'[18] The Commissioners were not criticising Allsop's morals. It was his political (Royalist) and religious views which they found unpalatable.

Brassington also showed a lack of enthusiasm for the new Church government in another respect. Each parish was ordered to send one or more of its elders to the next highest Church body, called the Classical Assembly or Classis. This met at Wirksworth and had a general oversight of the presbyteries in its district, plus the duty of examining candidates for the ministry. The minute book of the Wirksworth Classis for 1651–1658 shows that it was largely composed of clergy – they always outnumbered the laymen, listed as 'others'. The Classis exhorted the village churches to send their representatives. There was no response from the congregation at Brassington. The village was included

Sir John Gell, from the portrait formerly at Hopton Hall.

in the programme laid down by the Classis. The *Directory of Public Worship*, which had been substituted for the prayer book in 1644, directed that there should be monthly fast days, which included lectures and sermons. At Wirksworth the fast was called the Classical Lecture and held on the second Thursday of the month, in different places. There were three at Brassington between 1651 and 1654. The curate at Brassington must have obeyed the instruction not to use the old prayer book, since the Classis meetings have no record of any penalties being inflicted on him for doing so. There were fines of £5 for the first offence, £10 for the second and a year's imprisonment for further offences – enough to deter even the scandalous Mr Allsop.

By the time of a religious census held in 1676[19], Brassington had 375 conformists (to the Church of England, which had been reorganised on the old pattern after the return of Charles II), no papists (Roman Catholics) and only five Nonconformists. The census, which at Brassington covered everyone, including children, shows that the village had emerged from the Commonwealth years, when there must have been more than five Nonconformists, since nonconformity flourished then, safely back in the Anglican fold. The vicar of Bray's way of looking at things was probably the popular way.

TITHE WARS

The Commissioners of 1650 reported that the tithes belonged to George Lawcocke of Nottingham. They amounted to £50 a year, and Lawcocke and his heirs recommended candidates for the curacy. This was a poor living, a point made in a note written into the bishop's transcripts of the parish register in December 1682:

> We have no tarrier nor glebe land to ye church but Charles Lacock improprietor hath for a long time out of ye tyth of ye tenth allowed ten pounds a year to ye curate of Brassington. Mr Lacock lives at Woodborow in ye County of Nottingham. William Hand Curat ibid Richard Buxton William Wayne Churchwardens.

While thus helping the parson, the tithes were nevertheless a continual source of irritation and conflict in the village. The records of one long-running dispute heard at the bishop's court in Lichfield have survived.[20] In 1636, 1637 and 1638 a succession of Brassington men gave their opinion as to whether or not they were liable to pay 'tenths' of hay and corn to John Gell and William Greatrax, claiming then to be the 'impropriatories'. John Buxton and Edward Lane described how much hay and corn was harvested from the fields called, respectively, the Bents (Buxton) and Middle Hills, Meadowlands and Shelbread (Lane) and claimed that no tithes were payable. Buxton was supported by the memories of one old man, Robert Wayne, and Gell and Greatrax by another, the miner William Briddon. Briddon described Gell and Greatrax as 'the true lawfull farmers propriatories occupiers and possessors of all and singular the tenthes of all manner of corne graine and hey growinge renewinge and increasinge within the village or hamlett of Brassington.' For their part, Gell and Greatorex declared that almost the only parts of John Buxton's 'allegation' which were true were his statements that he was the heir of Richard Buxton and that William Westerne was the heir of Thomas Westerne.

By the time of the 1667 dispute[21] the value of the tithes had fallen because of the considerable switch from arable to cattle farming – hay was less valuable than corn and there were fewer sheep so that the wool tithe was less. John Buxton had been dead about twenty-eight years and his successor Richard had let most of his land and now paid only 20s per annum for tithe hay, instead of the £5 he had paid when he farmed the land himself. Others in the dispute were Henry Travis (Trevis) and 'Widow' Lane. No one in the village was willing to get himself embroiled in the dispute by gathering in the tithe hay and there was in any case no barn to store it in. It was noted that the corn tithe 'cannot be let for above 20L this yeare out of wch the curate hath 10L Mr Milward 5L & ye vicar of Bradbourne 13s 4d.' This dispute continued for six years, and the correspondence between the impropriator, Phillip Lacock, his son Charles and a number of villagers, reveals the determination of Henry Travis in particular to avoid paying the tithe. Under the will of Robert Gale, the original lay proprietor after the dissolution of the religious houses, certain legacies, in addition to payments to the curate and the others, were to be paid annually. One of these beneficiaries advised Lacock in 1667 to appoint 'two able men' to collect the tithe hay and corn and, if they were opposed by the copyholders, to bring them to court on a charge of riot, and to raise the question of tithe during the court proceedings. The letter giving this advice, from the Vintners Co. in London, asks whether the copyholders wish to 'treade the poore under foote nowe the cittie of London lyes in ashes' (1666 was the year of the Great Fire). Clearly the £20 legacies left to the Vintners Co. and other bodies were spent on charity. Lacock took the advice and persuaded his tenant, a Hognaston yeoman called William Cockaigne, to prosecute the case for him.

Cockaigne's letters tell the story of his attempts to get witnesses to speak for him at court, of Henry Travis persuading the miners not to pay tithe and of how the villagers 'laugh you to scorne'. Cockaigne in fact found his witnesses, and they included Richard Buxton and Robert Westerne. Travis found four men to speak for him, but lost his case when two of them confessed before witnesses that he had bribed them. One man was promised £5, plus 20s more if Travis won, the other £4 and a coat. This was in 1669. A letter from Cockaigne the following year asked Lacock to come to Brassington to restore order, as Travis had managed to persuade the miners not to pay their tithes and was himself still refusing to pay hay and corn tithe, in spite of having lost his case at Derby Assizes.

Cockaigne finally gave up in 1672 but by then John Buxton, nephew of Richard, was acting for Lacock and wrote in June that he hoped to settle the dispute soon. Buxton bemoaned the expense of the case – 'This bearer can tell you that I want money, my Coate as well as my purse being rubbed threadbare this last sumer.' The dispute was settled and Lacock received his dues. Their value was best shown when they were exchanged for an allotment of common land at the enclosure of Brassington's wastes in 1808 – the current impropriator received 350 acres.[22]

The villagers' spiritual needs were met by ministers whose livelihoods were little better at the time than the poorest of them. They included George Beresford, 'late minister of Brassington', who owned one bed, a number of unidentified 'small things', and a number of books. Beresford's richer parishioners had their own pews in his church, and there are wills as late as the nineteenth century bequeathing them. In 1710 it was reported that the Buxton arms were carved on the Buxton pew[23] and it is apparent that social distinctions were maintained inside as well as outside the church.

EDUCATION

George Beresford probably used some of his books to teach the children of the village. An Act of 1536 had made it compulsory for parishes to give reading lessons and religious instruction to all children – an essential measure if the people were to use the newly-translated Bibles and prayer books introduced after the break with Rome. The instructor was almost always the village parson. That there was education in Brassington is seen from the horn books, abacuses, grammar and 'accidenses' (word books) in the Kempes' shop. The horn book found in recent years behind a skirting board in the Tudor House, in this case made from the local metal, lead, is a survivor from its time as Thomas Westerne's or Ralph Marple's inn, and was perhaps bought in the Kempes' shop.

Only a few villagers, however, learned enough letters to enable them to write their names. These few, as might be expected, were the handful of gentlemen and the more prosperous farmers. The restricted group of men who were repeatedly called upon to appraise their neighbours' goods for probate – the Knowleses, Westernes, Buxtons, Wilcocks and the rest – usually signed the inventories of the goods they had appraised. There were exceptions even in this class, Richard Knowles in 1673 and Ralph Marple in 1695, for instance, signing their wills with their marks, and most of the women were illiterate. Anne Buxton, Anne Lane, Joan Knowles and Joan Goodwyn in the late 1600s all signed their wills with a mark. Even the rich families saw no reason to teach their daughters to read and write.

Among the husbandmen, labourers, craftsmen and miners, there was no one who could sign his name. Anthony Steeple, William Charlton, and Thomas Torr, husbandmen, all made their marks for signatures when they appraised John Charlton's goods in 1640. The inventory had been written for them by a professional clerk. Similarly, when Richard Jackson's goods were appraised in 1637, Thomas Gretton, Thomas Slacke, Jackson's brother-in-law and John Lee could only make their marks at the bottom of the inventory. A fourth appraiser, German Buxton, signed his name, and it was probably he who wrote out the inventory. When Henry Spencer the blacksmith made his will in 1632 he made his mark, as did two of his witnesses. The third witness was William Westerne, who signed. The Westerne family produced the only literate member of a trio who appraised the goods of the husbandman Edward Toples in 1632: this was William's brother, Richard. The situation had not changed by the end of the century. Henry Allsop, blacksmith (1673), Stephen Barton, miner (1696), Frances Briddon, miner's widow (1691), Thomas Buckley, miner (1696), Jonathan Hill, baker (1689), John Mellor, husbandman (1700), George Steeple, miner (1668) and John Tomlinson, miner (1691) all signed their wills with a mark, as did almost all their witnesses and appraisers. Clearly the 1536 Act had been ineffective in Brassington. The horn books must have been fastened to the belts of very few of the village children.

REFERENCES

1, 2 & 23 Glover, 1829
3 MLSL BM Add MS 66909 f8
4 Ince, 1871-72
5 DRO Q/Rl c5
6 DRO D161 B/6/52
7 NA E 317 Derb 9
8 DRO QSB 100
9 DRO QS IV crime box 51b
10 DRO QSB 205
11 DRO QSB 264
12 Hey, 1980
13 DRO QSB 914
14 Slack, 1987
15 MLSL BM Add MS 6671 f47
16 British Library, *Thomason Tracts*, E242 (6) (quoted in Wood, 1997)
17 Cox, 1880
18 Cox, 1877
19 Cox, 1885
20 LJRO B/C/5 1635, 1637
21 DLSL Gale Bequest bundle 3
22 DRO Q/R1 c5

CHAPTER 7

GROOVERS AND THEIR BARMASTERS

There had 'always' been lead mining in the villages of the Wirksworth wapentake, and long before the wapentake existed the natives whom the Romans set to work mining the veins on Carsington Pasture and Brassington Moor were likely already to have been expert at it. There is evidence of lead mining during Saxon times, and in 1289 Edward I, as part of a survey of the Crown possessions known as the *Quo Warranto*, ratified a set of rules and customs for the industry which were already ancient. To the men and women, of Brassington, lead mining was a natural activity and always had been. Every day, for centuries, there had been men, women and children getting lead from the limestone under the thin soil of Carsington Pasture. Lead mining died when the great mine at Millclose, near Darley Bridge, stopped producing in 1939, and the sort of mining carried on at Brassington had been moribund for a century by then, killed by cheaper imported lead. Its dying took a long time – the habits of centuries were not easily thrown off and men whose fathers and grandfathers had had their little mines hankered after mines of their own. The forms and ceremonies of the industry are still preserved in a Derbyshire version of Olde Englande, and the barmaster and his court still meet in the Barmote Hall at Wirksworth.

MINING METHODS

The remains of the old mining industry consist of shafts and rakes, hillocks of waste material, and the ruins of limestone buildings, which include the miners' 'coes', or cabins, in which they stored working clothes, candles, tools etc. Lead veins were often close to the surface and mined by means of trenches while this was possible – the origin of the word 'groove', used for 'mine' up to and beyond the seventeenth century. The early trenches, 'rakes', are often punctuated by shafts sunk later, when the vein had become too deep to be reached in any other way. These are within a few yards of each other, reflecting the difficulty the miners had in ventilating their workings – a new shaft let in more air.

The miners' work was extremely arduous and dangerous and only to be undertaken when the price of lead was right. The gunpowder in Anthony Kempe's shop may have been used in the mines but in the sixteenth and probably for most of the seventeenth century the main tools and methods were more ancient, hardly different from those used in Roman times. The shafts and trenches were sunk and the galleries cut with picks and shovels, hammers and wedges as the veins were followed from the surface. The picks were straight, enabling the miners to hammer the pointed ends into the rock strata. Wedges were used in the same way. Using these methods the miners' advance through the rock was hard and slow. Before blasting began to be used in the seventeenth century, pick and wedge driving was sometimes varied by 'fire setting' or 'lime blasting'. In the former a fire was lit against the rock and the hot rock doused with cold water, causing it to split; lime blasting was achieved by boring a hole in the limestone and filling it

Left: A mining rake on the eastern slope overlooking the village.

Below: The remains of a washing buddle at Perseverance mine on Carsington Pasture.

with quick-lime. The hole was stopped with a wooden bung with a small hole drilled through it. Water was poured through the hole on to the quicklime and the resultant chemical reaction split the rock. Fire setting added suffocation to the dangers of being crushed by rock falls, falling down the shaft or being slowly poisoned by the lead, and was only done at the end of a working day. When black gunpowder was introduced it was poured into a borehole, stopped with a plug of clay. A hole was punched through the clay, a straw fuse inserted, and another hazard brought into the mines.

While they were hacking out ore from a vein the miners piled the waste rock on the sides. In shafts, the waste, or 'deads', was piled on wooden platforms built across the vein, while work continued under and to one side of them. The miners climbed up and down shafts, often sunk within the coe, by ladders, by 'stemples' (wooden steps built into shaft walls) or, very precariously, by footholds. Lead ore was brought to the surface up 'winding shafts' by windlasses ('stowes'), though the ancient stowes were replaced during the nineteenth century in some mines by horse gins, where the power for lifting the ore was supplied by a horse trudging in a circle. In two mines, Great Rake and Nickalum, there were engines to do the job in these later days. Hand winding was often done by the miners' wives and children, who were also responsible for washing and dressing the ore. Dressing consisted of crushing the ore with heavy, flat-headed hammers ('buckers'), though horses were sometimes used in this process too in the nineteenth century. There is a site on the slope below Nickalum mine, where a horse pulled a roller over a stone circle on which the ore-bearing rock was crushed. Washing consisted of separating the heavier fragments with high lead content from lighter ones by either plunging the crushed ore in a sieve, or 'jig', into a trough of water or by washing water over it in a trough sunk into the ground – a 'buddle'. There is a stone buddling trough at Great Rake, and at Perseverance mine the remains of a buddle have left a shallow depression on the slope near the mine hillock.

This was the nature of lead mining in Brassington. The mines were always there and men in every generation learned the skills to enable them to take advantage of the trade's unique laws and customs. Mining was an adventure, and while prospecting was always a gamble, it had an overwhelming attraction to men who would otherwise have been wholly dependent on farm work. Compared with the life-long drudgery of labouring in the fields in the certain knowledge that the master would never pay them more than the minimum needed to survive, that they could be laid off in bad times, that they would be unlikely to save anything for their old age, and that their life's work would leave them bent and exhausted, mining offered independence and hope. The 'poorer sorte' preferred to 'labour in the lead groundes' because there they were their own masters and because from the middle of the seventeenth century, for about 150 years, mining was profitable enough to pay the rent of the few acres which would feed a few sheep, cattle or pigs.

THE RULES

Edward I's *Quo Warranto* confirmed the peculiar rules of lead prospecting, and they persisted into the nineteenth century. Lead was the king's property and the rules reflected this fact – they overrode all other rules of land use and made it easy for individuals to get and sell any ore they found. In the medieval period and earlier, when the amount of cultivated land was small, the miners were the men who made profit from the moorland where the lead was usually found. The rules devised for this situation were retained long after the face of the countryside had been transformed. Though the system, not surprisingly, guaranteed that a large slice of any profit from the mines went to the king or to his 'farmer', the entrepreneur who had leased or been granted the mineral rights, the rules also allowed miners to prospect anywhere except under highways, churchyards or orchards, to make roads to carry their ore away, and to use water, including streams where there were any, to wash it. This licence to despoil farmland always caused conflict and the miners could always rely on the ancient rules. In the seventeenth century the conflict can be seen in the words of the custumal of 1620: 'by reason of the myners they are troubled with a desase among theire cattell called the Belland wherewith theire cattell being once infected are uncurable.' The copyholders complained of the 'great losse to the owners' caused by the activities of the lead miners and the same feelings were being expressed fifty years later when the manor-court laid down penalties against miners who damaged fences, left shafts uncovered or took water from the village wells to wash their ore.

The custumal pointed out that even when a copyholder himself mined his own land, 'his Matie or his farmer' took every thirteenth dish, a standard measure of about sixty pounds. This was the 'lot', a tax accepted by the miners as being sanctioned by ancient custom and a fair payment for their rights. They paid a second tax, called 'cope', of 4d or 6d in return for the right to sell their ore to whichever smelter they wished. A third tax, the 'tithe', was resented and avoided wherever possible. The tithe was a tenth of the miners' ore. Before the Reformation, it was used for the upkeep of the Church and to ensure that prayers were said for the miners. Once it passed into lay hands the miners of the Wirksworth wapentake saw no reason to pay it. There was bitter conflict in the seventeenth century with John Gell of Hopton over his ownership of the tithes of Bakewell, Tideswell and Hope. There survives a summons in 1627 to the Barmote Court to meet 'at Brassington... upon munday being the xiith day of November next comeing by nyne of the cloke in the morning att the house of William Westerne gent', to hear evidence in a dispute between Gell and a group of miners.[1] During a later case, about 1650, Gell summarised his argument:

> Whereas the miners say that they ar not able to work & paye: wee answer, that they have wrought & payd this 400 years & therefore no doubt so may they do still: & that al there neighbours at least 12 prishes do pay duty & yet worke very gladly & that but 3 that denye & that the miners in other Cuntryes paye much more out of their oare than wee do & have not halfe the privelidges that we have, & yet worke in great number & gladlye.[2]

It was in reply to one of several petitions from Derbyshire miners for relief from poverty caused by excessive taxes that Charles I offered to remit tithes for every miner who joined his army at Nottingham.[3]

The nature of land tenure in Brassington, which had developed over the centuries as an open village with no supreme lord of the manor, meant that the village miners were spared the necessity of protecting their rights against a powerful estate owner determined to run the mines for his own profit. In some villages the rights were vigorously contested by estate owners who claimed full ownership of the mines and used legal and physical means to end free-mining practices. In nearby Elton, for instance, the lord of the manor, Francis Foljambe, prevented the operation of the customary laws and employed labourers in the Elton mines and as carters to take the ore to his own smelter.[4] Elton was one of the liberties of the Wirksworth wapentake and the barmaster opposed Foljambe in the duchy court and, more directly, by physical means. The court ruled against Foljambe, but this was one battle in a long war.[5] The Brassington men kept their independence.

BARMASTERS

The sales of ore to the smelters were presided over by the barmaster for the mining 'liberty'. He measured the ore, collecting every thirteenth dish in payment of lot, and the remainder was sold in loads and dishes (nine dishes per load), for which the barmaster collected cope from the ore buyers. He was appointed by the owner of the mineral rights and ran the industry. An agreement between Gilbert, 7th Earl of Shrewsbury, the current farmer, and a barmaster, John Shore, survives from 1597.[6] Shore was to collect the lot and cope, pay the queen's tax of £72, and share any profits from the mines. The Brassington mines came under the overall jurisdiction of the Wirksworth barmaster, with a deputy barmaster for the Brassington liberty. A survey by a Parliamentary Commission in 1650[7], when the victors in the Civil War were carrying out an audit of the assets of the Duchy of Lancaster, defined the boundaries of the liberties, the duties and perquisites of the barmasters, the customs and procedures of the mines and the current income from the Wirksworth wapentake and from each liberty. Brassington liberty, presided over by William Blackwell (Blackwall), was 'lymited & confined within the Towne Feilds of Brassington Bradboume Tissington Allsopp in le Dale Eaton Mapleton Thorpe Bentley and Kniveton, extending itself in & through-out ye same & every of them.' At that time the farmer held his rights on a thirty-one-year lease, still paying £72 for the lot and cope, plus a further 26s 8d for the office of barmaster. The deputy- or under-barmasters were either appointed by him or elected by the miners, and were local gentry such as Trevis and Blackwell. They had 'noe Fee or Sallary certaine Butt some proffitts att ye Measuring of ye oare, And other pquisitts hee hath by making of Arrests And serving att the Barremoot Court.' Barmasters did well enough from their period of office to attract the likes of Henry Trevis and William Blackwell. They were paid by the miners for measuring their ore, at a rate which in the eighteenth and nineteenth centuries was 3d per load (nine dishes)[8]. An 'arrest', in mining terms, was the barmaster's suspension of operations at a mine whose ownership was in dispute. The barmaster adjudicated and the mine remained arrested until the matter was settled.

The barmaster's duties included the 'gift' of mines. When a miner struck ore he staked his claim by presenting a dish of it to the barmaster, who recorded in a book that such-and-such-a mine had been 'freed' by the miner. The barmaster then gave him title to his new mine and with it permission to extract lead from two 'founder meers' of ground. A meer in the Wirksworth field was 29 yards long – there was no restriction on width or depth. A third meer, called by the commissioners the 'primme gappe' and by later miners the king's or lord's meer, was the property of the owner of the mineral rights and the miner could either buy it, at a price fixed by the barmaster, or work through it and present the ore from it to the barmaster. The miner could then free as many further meers as he wished. These were called 'taker meers'.

It is easy to imagine the power and influence which this system conferred on the barmaster in a village like Brassington, where most of the people were more or less dependent on mining for their living. It is also apparent that the 'pquisitts' noted by the commissioners were likely to amount to a respectable sum. The barmaster had a further power over the miners in the authority to take away the ownership of any mine which had not been worked for three weeks, except in the case of flooding or bad ventilation. He gave the owner three weeks' notice, marking each week by a cut on the mine's stowes, a process known as 'nicking'. There cannot be many modern thieves who know where that term comes from. Powerful though the barmaster was, there was a rough democracy in the mines. As the commissioners pointed out, the barmasters were sometimes elected by the miners of the liberty, and they were in any case assisted in their duties by a jury of twenty-four miners, meeting as the 'Barmote Court'. Twelve of the jurors acted as small courts in such cases as mine deaths, where the decision as to the cause of a miner's death was made not by the coroner but by the barmaster and his jury. It needs to be stressed that the mines were being run in the 1600s more or less as they had been in the time of Edward I'. While the authority of the manor-court and reeve had diminished, that of the barmaster and the Barmote Court remained, though discipline and order were maintained by fines rather than by

The Barmote Hall at Wirksworth.

the more physical methods of earlier centuries. Then, for instance, a persistent thief would have his hand nailed to his stowes, with the option of staying there until he died, or of cutting off his own hand. To make this a real choice, the barmaster would leave him a knife.

Surviving accounts for the dinners served to the Barmote Court at its meetings in the 1660s show a fine regard for the three ranks in the industry.[9] A 'bill for the Charge of the Barmot Court' lists four 'gentellmens dinars' at 12d each, fourteen at the barmasters' table at 10d and thirty-two at the '24 table' at 8d. There was beer and tobacco at £1 2s 6d and wine, presumably for the four gentlemen, at 8s. Notwithstanding this innkeeper's grading system, the local gentry with interests in the mines took care that they were themselves members of the '24'. A jury which issued a list of the mining laws in 1655, for instance, included Richard Buxton and Anthony Gell.[10] There were fifty-three laws in this list and the importance attached to them is seen in a note which the Barmote jurymen added: 'We say that theis articles ought to bee read and published againe the next Barmote Court in open Court that all men may take notice thereof'.

Examples of the system in action can be taken from any period. A document in the Gell collection, dating from January 1660, describes how four men of 'the 24 being this day called to a certain grove of John Mousley lying & being on Cromford More', found that Mousley and his partners were trespassing. The four jurymen received a shilling each for carrying out this duty. More than a century later Edward Ashton, then barmaster for Brassington, described the process of arresting a disputed mine:

> …when an arrest is made and executed by any of the deputy barmasters within the said wapentake and notice given to the party arrested… then such arrest cannot by any mineral law or custom whatsoever… be allowed discharged without trial of a small Barmote Court before the steward and a jury of 12 men to be impanelled for that purpose unless both parties in case of agreement shall by mutual consent agree to the same.[11]

'T.W.' was the miner killed in Throstle Nest mine, on Brassington Pasture, in the year after he carved this boulder nearby.

This eighteenth-century barmaster was restating the ancient custom, illustrating that his predecessors' wish that the rules 'ought to bee read and published againe' was carried out throughout the industry's history. An eighteenth-century textbook, the *Miner's Guide*, by W. Hardy, includes examples of the operation of the rules. In the second edition, published in 1762, a Brassington case is quoted: Ashton is asked to arrest Calveshead mine during a dispute between Thomas Scattergood and partners and William Turner and William Blackwall and their partners. There is also a statement by a Brassington jury who had been assembled by Ashton to establish the cause of a death in Throstle Nest mine on Brassington Pasture: 'it appears that a large stone fell upon him out of the roof, and it is our opinions the said stone was the cause of his death.' This is signed by twelve Brassington miners.

The system was notoriously subverted in the 1630s by the king's attorney general, who plotted to dispossess the holders of mines in an area at Wirksworth called the Dovegang.[12] The Dovegang had been mined to the water level and the attorney, Sir Robert Heath, had engaged Sir Cornelius Vermuyden, a Dutch engineer who had earned his knighthood by draining the East Anglian fens, to drive a sough, or drain, to dewater the mines. Vermuyden brought a fellow Dutchman, Conradus Molenus, to apply pressure to remove the owners of the mines and Heath made it legal by deceiving a Barmote jury chaired by an associate installed as temporary barmaster. There are the records of three meetings of the '*Curia Magna Barmoti*' at Wirksworth in 1630.[13] Among the ordinary business was a warning to miners that if they did not attend to their mines they would lose them, an instruction to Thomas Hawkesworth and Joseph Bond not to undermine Thomas Woodiwiss's barn, on pain of a £5 fine, and an instruction to John Cadman 'and his grovefellowes' to improve the drainage at their Dunbalkes mine, also with a £5 fine if they failed to act. The main business at these meetings was to observe the legal preliminaries to Heath's taking over the Dovegang and setting Vermuyden to work. Twenty-eight men and women 'who have or prtended to have in lawfull possession some meers of ground or rakes lieing in the wai of drayning or soughing the Dovegang' were ordered to attend 'to defend and maynteine the same'. All but six gentlemen landowners ignored the summons and were dispossessed. Molenus was to become a servant of John Gell and to command his infantry during the Civil War, with the rank of major. After the war, Molenus leased one of Gell's smelters in Wirksworth and bought lead ore from the area's mining liberties, including Brassington. The Brassington ore accounts refer to him as Major Molenus, or simply 'Major'.

The miners sometimes settled disputes without recourse to the Barmote. In 1797 four men, Christopher Slack of Brassington and Matthew Bacon, Joseph Beardsley and Richard Batterley of Carsington, were indicted at the Derbyshire Quarter Sessions for assaulting Joseph Allsop and 'wilfully forcing a large quantity of Stones and Earth upon him in a Mine at Carsington which much wounded and nearly suffocated him.' For this drastic action the four were let off lightly and discharged under sureties of good behaviour entered into by Slack's brother and friends of the others.[14]

THE TRADE

The lead trade waxed and waned. An analysis of prices during the sixteenth century shows a rise from 5s per load in 1540 to 10s in 1588.[15] An ore purchase book for the year 1541 to September 1542 shows that the total production of the Brassington mines was 385.5 dishes (forty-two loads, five and a half dishes, or about 10.5 tons), a small amount when compared to the figures achieved during the following two centuries. This analysis also includes the information that in 1541-1542 there had been thirty 'reckonings', the occasions on which ore was sold, and that they had occurred in nine months of the year – only in October, January and April had there been no ore sales, indication that there was no close season for mining. The villagers worked the grooves whenever there was lead to be got.

After the busy period presided over by the Earl of Shrewsbury came a period of low prices and hard times for the miners, made worse for them by taxes imposed by successive monarchs. The miners petitioned on several occasions to have these extra taxes removed, and a petition of 1642[16] gives a convincing picture not only of their grievances but also of the crucial importance of mining to their livelihoods. The fact that such a document as this petition, written on a very long roll of parchment, listing several thousand names, could be compiled by the miners, suggests they were a well organised, coherent community. The miners had a trade, a 'mystery' of their own and were conscious of their unique skills and of their value.

The petition, organised by a lead merchant called Lionel Tynley, was addressed: 'To the right honorable the knts cittizens and burgesses of the Comons house of Parliament. The humble peticon of twentie thousand myners whose names are hereunto annexed Inhabitants of Darbieshire on the behalfe of themselves and divers others'. The number 20,000 was probably not an exaggeration, though they were not all miners – there were mine owners and shareholders too. A contemporary, though unattributed, statement quoted in the *Victoria County History* gives 4,000 as the number of miners. Their complaint was that, in addition to lot, cope and tithe, extra impositions had been made by Queen Elizabeth, James I and Charles I. The miners were content to pay the ancient dues, knowing that 'all ye rest of ye lead ore is their owne both by lawe and ye custome of ye myne, they are discharged of all other payments for ye same.' After outlining the ancient dues and customs, and stressing the importance of lead and its 'profitt and comoditye to ye whole comonalty besides ye continuall mayntenance and daily ymployment of many thousands in & about these mynes', the petition came to the point.

Elizabeth had added a tax of 8s 'upon evrye fodder or tunne' of smelted lead, and 'the myners did beare it patiently.' However, there had been further taxes of 20s imposed by both James I and Charles I, at a time when the price of lead had fallen – 'When lead hath been at fowerteene pounds the fodder, the myners payd no other impcsitions but only the forsaid eight shillings And now lead is under tenne pounds a fodder yet they are forced to pay the other forty shillings also.' The petitioners foresaw 'the utter undoinge of them their wyves and children' if the 40s tax were not abolished. The names of the petitioners are listed under villages and the Brassington men are listed with those of Parwich, Carsington, Tissington and one which is now illegible. Each miner's name is accompanied by the number of his dependents, to ensure that Parliament would be in no doubt of what hung on its decision. There is a grand total of 474, which includes 142 'poore hirelings and cavers' with their families, who are not named – cavers were men and women who picked ore from mine waste heaps and the barmaster's records during the 1650s show that they made a significant contribution to the production in Brassington liberty. The hirelings were those employed as waged labourers – the independent miners were at pains to maintain their status and when they worked in other men's mines they worked in groups, carrying out negotiated contracts. Of the named miners, twenty-six have Brassington names and they and their dependents come to 124, probably about a third of the village's population.

The petition pointed out that in addition to the miners there were large numbers in supporting trades. There were 1,000 men constructing the Dovegang Sough, 'when it is in worke', 1,000 smelters, 2,020 'gaggers' or jaggers, 'carrying oare from ye mynes to the smilting milnes', and another 2,000 who carried smelted lead from the mills. There were 500 woodcutters, 'wich makes coales for the smilting milnes' – smelting at that time was accomplished by kiln-dried wood. There were 150 men making 'corves', wooden sledges used for hauling the ore underground, 600 woodcutters who cut mine timber and another 1,000 carrying corves and mine timber. Mines were lit by candles and there were 200 chandlers making them. There were 100 rope-makers and 150 smiths and 'naylers'. No doubt Henry Spencer forged tools for the miners in his smithy in Brassington. The joiner John Allsop perhaps made mine doors. The Kempes' shop sold candles and no doubt the miners bought theirs there. The figures paint a picture of an industry which gave a livelihood either directly or indirectly to a sizeable proportion of the village population.

The slump in prices described in the petition can be seen in the inventory of one of the Westerne family into whose hands the lead trade had passed by the 1620s or earlier. This was William Westerne. In 1635 he had thirty-two loads, six dishes of ore, priced at 19s per load. This was almost twice the 1588 price, but should be seen against the great inflation of the sixteenth century. Westerne also had '19 small peeces of lead sould after' at £4 13s 4d per fother, a low price even when compared with prices regarded as ruinous by the petitioners – under £10. The difficulties caused to the miners by low prices and high taxation seem to have been fairly short-lived, to judge by the output of the mines and the evidence of modest prosperity at the end of the century. A survivor from Henry Trevis' accounts shows a total of sixty-seven loads measured at Brassington during the months of July to September 1639.[17] In 1655[18], the only year for which all the monthly totals have survived, in the Gell papers, William Blackwall measured 687 loads of 'grove' or mined ore and 136 loads of 'caved' ore – ore reclaimed from waste heaps. The relatively healthy state of mining at Brassington was shown by the annual value of £66 which was placed on the liberty's income by the 1655 survey. This comprised lot, cope:'And allsoe all that the future & casuall Benefitt by forfeitures for felony & by concealing of oare or otherwise together with such Meeres of Ground and Primme Gaps wch by finding a New Veine of oare may by possibilities accrew to the farmers of the Lott & Copp or their Barrmrs'. The second highest annual value among the liberties of the Wirksworth wapentake was £66, after Matlock, Wensley and Snitterton's £80. For the eighteenth century there are few records surviving for Brassington before the 1790s. The barmasters' books recording gifts and sales of mines and the reckoning books, listing the mines, their owners and their output, have all been lost, with the sole exception of the reckoning book of Thomas Slack, barmaster from 1792 until his death in 1804.[19] This book, put to later use by his son and great-grandson for the accounts they kept for their pub, the Miners Arms, has been preserved by his descendants.

A record of a different kind is the account given by Daniel Defoe, in his book *A Tour Through the Whole Island of Great Britain*, of his encounters with a Brassington miner and with a miner's family. Defoe, on his way from Wirksworth to Buxton sometime in the mid-1720s, was looking out for what he had been told at Wirksworth was a giant's tomb when he came across a woman living with her five children in a cave in a hillside. Her description of the giant's tomb as a flat stone 16 or 17ft long sounds like the capstone of a chamber grave and this, together with Defoe's description of her cave home, makes it likely that the place was Harborough Rocks. The family had a small enclosure with a crop of barley and there were pigs and a 'little lean cow' about the entrance to the cave. The cave itself was clean, divided into three rooms by means of curtains, and equipped with shelves, earthenware pots and pewter and brass vessels. The woman told Defoe that her husband was a miner and that he and his father before him had been born in the cave. He earned about 5d a day and she, when she could spare time from looking after her children, could earn another 3d a day washing ore. Defoe describes her as 'tall, well shaped, clean, and (for the place) a very well looking, comely woman'. Her children were healthy and he marvelled that such apparent wellbeing could be achieved on a daily income of 8d. He was sufficiently impressed to give her a present of 'a little lump of money, though the sum was not great, being at most something within a crown (5s)' and, abandoning his search for the giant's grave, he followed the woman's directions to where there were a number of mines.

Defoe and his party were examining the shafts when 'we were agreeably surprised with seeing a hand, and then an arm, and quickly after a head, thrust up out of the very groove we were looking at.' When the miner – 'this subterranean creature' – had climbed out of the shaft, Defoe questioned him and was struck both by his appearance and by the nature of his work. The miner, he said:

> …was a most uncouth spectacle; he was cloathed all in leather, had a cap of the same without brims… For his person, he was lean as a skeleton, pale as a dead corps, his hair and beard a deep black, his flesh lank, and, as we thought, something of the colour of the lead itself…

Defoe marvelled at the narrowness of the shaft the man had just climbed, until he examined the wooden stemples built into the shaft walls and realised that the narrowness made it easier for the miners to use their elbows as well as their hands and feet as they clambered to and from their work. Through an interpreter, since Defoe could not understand the miner's Derbyshire dialect, the miner told him that he was mining at sixty fathoms (360ft) and others at seventy-one and eighty-six fathoms. These others 'had a way out at the side of the hill', in mining terms an 'adit', which may mean that the mine was on the Carsington side of Carsington Pasture, where the hillside slopes steeply. There has been doubt about the accuracy of Defoe's account, but his closely-observed description of the climbing shaft and its stemples, of the miner's clothes, in particular the brimless leather forerunner of helmets, and of the fact that one miner had to climb to the top of the hill to descend to sixty fathoms while his partners walked into their much deeper working area by a horizontal tunnel, all have the sound of truth.

Defoe thanked God 'that we were not appointed to get our bread thus, one hundred and fifty yards under ground, or in a hole as deep in the earth as the cross upon St Paul's cupulo is high out of it.' He insisted on his companions each buying a piece of ore, and gave the miner 2s for it. This was more than he could earn in three days' work. Defoe

The entrance to the Harborough Cave inhabited for many centuries and probably the home of the family visited by Daniel Defoe.

finished his story by describing how the miner went forthwith to an alehouse in the village to 'melt some of (the 2s) into good Pale Derby', and how he bought the man a drink and persuaded him to take the 2s back home to his family 'which they told us lived hard by.'

These Brassington people are anonymous, but Defoe's vivid narrative illustrates the poor, but far from poverty-stricken lives of the miners during a good time for the industry. It also makes it clear that the miners were able to maintain their independence and, while the cave home was presumably unique, the family's barley field, pigs and cow were typical. The picture is confirmed by miners' probate documents during the first half of the century.

The only detailed record of the Brassington mines for the eighteenth century, the reckoning book covering the years 1792 to 1803, shows the beginning of the decline in the industry. There are figures for nine months of 1792, during which time 848 loads of ore were sold. The figure was 929 during 1793; a higher output than had been achieved in William Blackwall's time, a century and a half earlier, and probably typical of the eighteenth century. In 1801 the mines produced 717 loads, and a reckoning book for the 1820s[20] shows a further decline to around 600. There were corresponding falls in the number of mines from 132 to eighty-two between the 1790s and the 1820s, and of named mine 'proprietors' from ninety-six to eighty-six.

These detailed figures show that most of the mines at Brassington were very small ones indeed. A later barmaster characterised them as, 'Poor men's mines. That is capable of supporting 2 or 3 working men.'[21] Most, in fact, would not have supported one miner, and were clearly providing additional income to what was earned from seasonal work for the farmers or from the miners' own small holdings. To put figures to this, thirty-six of the eighty-two mines listed in the 1820s produced less than five loads for the whole period, and a further twelve between six and ten loads. Only thirty-four of the mines listed in the 1790s had any ore measured in the 1820s.

However, there were always a few mines producing respectable amounts, enough to keep a family, and in every period Brassington seems to have had at least one big mine. In both the 1790s and the 1820s and, from other evidence, throughout the intervening years and up to the middle of the nineteenth century, the Balldmeer mine was producing very

The miners' coe at Breck Knoll mine, on Carsington Pasture. This coe has a fireplace, with a chimney still standing, a floor of limestone flags and cubby-holes on each side of the fireplace for tools, tobacco, food etc.

large amounts – a quarter of all production in the 1790s and a slightly smaller proportion in the 1820s. Its product was mostly 'belland' or lead dust, which brought a much lower price than ore, but the mine was nevertheless profitable enough to make a good living for its proprietors, the Fearn family. Even more productive, though for a shorter period, was Matthew Bacon's Children's Fortune mine, on the slopes above Carsington: Bacon was the landlord at the Miners Arms there. Children's Fortune produced 1,316 loads between 1797 and 1802. Mines which at different periods produced enough ore to keep rather more than '2 or 3 workingmen' included Wayne's Dream (184 loads in 1818 and 1819), Perseverance (219 loads between 1821 and 1826) and Providence (142 loads between 1822 and 1825). Great Rake mine, which by the 1820s had been mined for at least 170 years (it is listed among the mines at which William Blackwall had measured ore in the 1650s), produced 141 loads in the four years between 1822 and 1826.

From 1808, when the barmaster's son Thomas, also barmaster, became landlord of the Miners Arms, the ore accounts are replaced by the accounts run up by the landlord's pub customers. The commodity most mentioned was, of course, drink, and, since this landlord was the Brassington barmaster, it was more often supplied to miners than to any other category among his customers. The most frequent entry is for ale supplied to a group of miners and charged to the named proprietor of the group: '1816 Jany 1 ale two gags (jags) with Ellett & Co 0-7-11.' The ale was sometimes taken to the mine, sometimes drunk in the pub: '1815 May 20 ale to mine by Taylor', or, '1826 Aug 15 ale by Benjn Tomlinson and in House by Pakeman 0-9-5.'

Starting a new mine was a great occasion, celebrated with music and drink: 'ale & fiddling' (5s 6d) at the opening of Wayne's Dream and, at another mine, 'ale at giveing a Mine & my trouble 0-17-0.' The 'trouble' was the barmaster's fee for registering the title to the mine, put in another entry at 3s. While the miners usually drank ale, there are also entries in the accounts for gin, brandy and rum, not to mention 'rum & watter'– this Derbyshire pronunciation of 'water' was still to be heard in the twentieth century. The accounts rarely mention what was eaten at the pub, but Matthew Bacon and his team of miners must have relished their evening there on 4 September 1813, when they had 'ales and Cakes after' at 2s 4d.

Apart from the hundreds of entries for supplying ale, which seems to have been an essential lubricant for work at the time, the pub accounts make it clear that the land-lord supplied endless credit for the mining operations themselves. These were small mines and needed little capital. The miners were poor men, however, and the credit was needed. The barmaster's 'trouble' included giving and freeing and, most important, measuring. The measuring fee was given later in the century as 3d per load, confirmed by comparing the amounts measured in the 1820-1826 measuring book with the debts incurred at the pub. These could be sizeable amounts – the barmaster's uncle, Job Slack, and his son, Thomas, who worked Wayne's Dream during its short productive period, had forty-six loads measured on 13 July 1818 and thirty-two more on the 19th, for which the landlord entered debts of 11s 6d and 8s respectively. He supplied the wooden dishes in which the ore was measured – 'To a meer dish at Flaxpiece 0-6-2' – the stowes or windlasses which signified ownership – 'Stowes to Balldmeer 0-2-0' – and gave credit for expenses incurred when the jury met at the pub to decide a dispute – 'To ale Eating & pd the Twenty four and other expences the mine acct 1-14-0.'

The measuring books and the Miners Arms accounts give a deeper and more detailed picture of the miners' work than any of the earlier records. They make it clear that, at Brassington, things were still being done in the old ways – the miners still worked in partnerships or 'companies', as the barmaster called them, they still marked their titles by placing their stowes on the workings, their ore was still measured in wooden 'meer dishes', and the industry was still governed by miners' juries, meeting at the barmaster's pub. Mining was still an adventure, and new mines were given ceremonial openings. One entry in the pub accounts reminds us that the Brassington men were working in a dry area – the landlord supplied cartloads of water.

In the nineteenth century, near the end of the long mining era, local newspapers reported anything sensational, and *The Derby Mercury* and *Derby Reporter* both had accounts in 1837 of the discovery of a newborn child found dead in Fleelow mine. This grim item was unique, other stories more typical of the events which had characterised the industry for centuries were disputes at the Prince Albert and Victoria mines reported in 1847 and 1848, and at Balld Meer and Harborough Rake in 1863. In 1862 the *Derby Reporter* gave an account of a speech at a meeting of the Great Barmote Court by the owner of the Old Brassington (Nickalum) mine:

He had the pleasure of announcing to his friends that by perseverance for upwards of twenty years, they had at last found the long sought for treasure, which he hoped would be prosperous, and they should be able to continue employing, as they are at the present time, upwards of 100 men at one mine in Brassington.

This report was in the issue of 2 May. A fortnight later the paper reported an accident at Old Brassington. Taking a break during a night shift, a group of miners had come up the shaft, settled in the candle-lit coe and lit their pipes. Also in the coe, amazingly, was a store of gunpowder in three-pound bags. One of the candles, or someone's pipe, ignited a bag, which blew the roof off the building and injured four of the men. One, William Orme of Cromford, 'was dreadfully burnt, his body being scorched and his face dread-fully disfigured. Everything calculated to allay his torments was done until he could be conveyed to his home... we are glad to hear upon enquiry that hopes are entertained of Orme's recovery.'

Relaxing, possibly at Golconda mine, about 1860. The owner, Matthew Bacon, with the bottle, is sitting on the mine's wooden buddle with Joseph Repton and his son, also Joseph. While Bacon is in working clothes, the Reptons are dressed in their best – one family tradition has it that they were returning from a meeting of the Barmote Court. The winding gear in the background was operated by horse-power.

The Miners Arms.

A contemporary book[22] described how the miners celebrated their annual holiday on 13 May. They decorated their coes with garlands 'and other rural decorations', ate a good dinner, listened to a band and joined in singing 'old songs peculiar to the miners'. Twenty years ago there were still memories in the village, passed on through the generations, of this late, most unusual, and last, period of high production. Perhaps a good way to end this chapter, with the end of lead mining very near in Brassington, is with one of them – a picture of the miners at Old Brassington, all 100 of them, celebrating their May holiday with beer and sandwiches in the field below the mine hillock.

REFERENCES

1 DRO D258 42/15
2 DRO D258 54/19ha
3 MLSL BM Add MS 6677 f47
4 Wood, 1999
5 DRO D258/30/16-24)
6 SCA Bacon/Frank collection 21148
7 Slack, 1988
8 Stokes, 1964
9 DRO D258 28/20d
10 DRO D258 28/20l
11 MLSL BM Add MS 6676-6686 fl125-127
12 Slack, 1994
13 NA DL30/54/669A
14 DRO QS
15 Kiernan, 1989
16 NA E101/280/18
17 DRO D258/58/24a
18 DRO D258M 28/20a
19 DRO D2629 zl/l 1
20 DRO D2629z
21 CHBC 9/100
22 Hackett, 1863

CHAPTER 8

A WORKADAY VILLAGE

A NEW WORLD

The great changes in landholding, agriculture, village government, religion and in the social patterns among the villagers, which together transformed the medieval village, were largely complete in Brassington by 1700. The villagers had long been free and able to sell their labour where they could, but they could no longer graze their cattle and sheep on the open fields in winter after the crops had been brought in. For grazing they paid those few of their neighbours whose tenure of most of the land was now exclusive. There was still the open moorland – 'the moors and wastes of Brassington' – but here too the grazing had become concentrated in a few hands and most of the villagers had to pay rent for the privilege of grazing their cattle and sheep there. It was still a vital part of the villagers' lives, however, this immemorial wasteland, still unfenced and criss-crossed with the paths to Elton, Ible, Winster, Wirksworth and Hopton. It must have seemed impossible to the villagers that they would ever be fenced out of this part of their territory, and yet the landowners with grazing rights there were intent on enclosure and would eventually accomplish it.

The crops of oats, wheat, barley, rye, beans and peas were much diminished and replaced by larger numbers of cattle and sheep, while food crops were brought in from places with richer soil than Brassington's thin covering. The manor-court still met regularly to carry out transfers of the copyhold fields in the former duchy-manor, but it had finally given up its role in village government by the eighteenth century. This role was played by the officers of the parish and by the Quarter Sessions of the magistrates' court in Derby, and one of their chief preoccupations was the relief of the poverty which the modern system had created. Modifying the new system were the ancient rules of mining, very important in Brassington during the industry's heyday in the seventeenth and eighteenth centuries, when it was worthwhile for the men of the village to take advantage of them and go prospecting. After the religious upheavals of the preceding two centuries, the eighteenth-century curates could attend to their small congregations, teach a few of the village boys their catechism and perhaps more, and baptise, marry and bury their parishioners. They had no competition until the arrival of Methodism in the nineteenth century, when there was a surge of enthusiasm for new faiths – enthusiasm which the parson would no doubt have found embarrassing if he had found it among the pews of the old church.

HARD TIMES

There is testimony from church and state of hard times in eighteenth-century Brassington. In 1727 William Bainbrigge of Woodborrow in Nottinghamshire, the then holder of the tithes of Brassington church, wrote to the bishop at Lichfield[1]. Bainbrigge had the privilege of nominating the curate at Brassington and his letter was a testimonial to a young man called Humphrey Simpkin. Simpkin had 'not been so happy as to have a University Education and therefore it may be not so well qualified as what I and your Lordship could wish', but Bainbrigge made it clear that in his opinion Brassington could expect nobody better: 'Consider the Necessity of the Church and that it is so small an income that scarcely anybody will accept of it.' It was a poor living in a poor village, and Simpkin was likely to be popular with the villagers because his services would be cheap: 'he hath hitherto Discharged well his duty in Teaching a School and been very Charitable in Teaching poor Boys Gratis which will without Doubt Render him very Acceptable to yt people who are ye Great part of them very poor.'

To this picture of a village of very poor people may be added a second characteristic noted in a later church document.[2] This was the 'primary visitation' of 1772, where the bishop posed a number of questions to the curate about the parish. One of them was 'what Families of note are there in it?' the answer was, 'No family of note in it.'

The last resident lord of the manor, Richard Buxton, died in 1724, leaving his 'moiety of the manor of Brassington' to his cousin William Newton. This was the old duchy-manor, and both Newton and the holder of the former Shrewsbury Manor, George Savile, were absentee landlords, as were their successors. Richard Buxton was a bachelor and the last of his family in Brassington. The last Westerne, Robert, had died in 1679; the Travis family had left the village. The Allsops, Briddons, Spencers, Waynes, Wilcocks and the rest of the families who lived in the village during the eighteenth century contained, as far as polite society was concerned, 'no family of note'.

The probate inventories which continued to be drawn up during the first half of the eighteenth century support the picture given by William Bainbrigge in his letter to the Bishop of Lichfield. Whereas there had been eleven inventories with values of more than £100 in the second half of the seventeenth century, 29 per cent of the total, there were only two in the next fifty years – 4 per cent of the whole. The contrast is in fact greater than these figures suggest, as there had been two inventories between £200 and £299, three between £300 and £399 and one of £632 in the seventeenth century, while the only one over £200 in the first half of the following century was John Buxton's £215 in 1703. Most of the inventories – about 60 per cent – were of under £30 in both periods. The difference lay in the figure of 34 per cent spread between £40 and £99 in the eighteenth century, compared to 14 per cent in the seventeenth. The village, judged by the inventories, had changed from a poor community with a few rich farmers and landowners to one which, while still poor and having no wealthy families at all, had considerably more families who were enjoying a limited and modest prosperity.

The lives of the eighteenth-century villagers are recorded in two survivals which are only preserved from the last years of the previous century – the parish registers of baptisms, marriages and burials and the churchyard gravestones. Some things are clear: most couples married in their mid-twenties and enough babies were conceived before marriage to make it likely that the working people of the village waited until they were making their own way and knew that they had started a family before they married. Most couples had large families, and many babies died. There are many burials of women who died soon after the birth of a baby. There were old people, including the occasional centenarian like Robert Johnson. He had been, like many others in the village in the eighteenth century, a 'hosier and worsted maker', and had enjoyed good health until a short time before he died in 1788, nineteen weeks after his hundredth birthday. Reporting his death on 20 November, *The Derby Mercury* added that his sister was still living in Brassington, aged ninety-six. The Johnsons were unusual. Few of the villagers reached three score and ten.

The village was, and always had been, a community in which cold and damp cottages, dirt wherever the people went, poor diet and untreatable disease made old age rare. The normality of early death persisted until quite recently and yet the twentieth-century innovations of comprehensive social and health services have made it very difficult for us to imagine what it was like for couples, the majority of whom saw half their children die, or for the mothers, whose every pregnancy was a hazard. By the second half of the century some villagers were using gravestones to express their feelings about life and death, with standard verses in the characteristic eighteenth-century tone – tough and philosophical, accepting the shortness of life, the inevitability of death and the remorseless progress of disease at a time when medicine was of little use. 'Short was my time/ Longer is my rest/ God called me hence/ He thought it best.'

PAUPERS

The parish registers put some figures to William Bainbrigge's statement that most of the people were poor. For most of the eighteenth century the curate noted in the burial register which of his dead parishioners had been paupers and it is apparent that many villagers who lived beyond working age were likely to need parish relief in their last years. During a three-year period between 1741 and 1744, there were nineteen paupers in a total of thirty-eight burials. In the next three years, nine out of twenty-eight people buried were paupers. At the end of the century there were four out of ten in 1786, five out of fourteen in 1787, and three out of fifteen in 1788. The number of entries with the word 'pauper' in the margin dwindled and there were none for 1794 onward. This did not signal a rise in prosperity among the people of the village; the curate had merely stopped recording which of them had been 'on the parish'. A second record of a villager's lapse into the care of the parish was made on the suite roll of the manor-court.[3] This medieval survival was maintained throughout the century, written on parchment and recording attendances at and absences from court meetings. Here again are the names with 'pauper' written against them and with a line drawn through the columns for the period when they had been paupers and had neither attended court nor paid the 2d fine for not attending.

The Act of Settlement of 1662 had been given a further cruel refinement in 1697 when one provision of a Poor Law enacted in that year instructed that a pauper and his wife and children 'shall wear upon the shoulder... a large roman P together with the first letter of the name of the parish.' In addition to being thus branded and humiliated, paupers earned their relief by hard labour on road mending for the parish or labouring for the village's farmers. There are examples of settlement certificates for those fortunate enough to be allowed to live where they wanted to and for the wretched families who were herded from one village to another.

Settlement certificates and removal orders were both printed documents, well known and understood by the 'poorer sort'. In a settlement certificate of 1762 for the blacksmith George Briddon[4] the Brassington churchwardens and overseers William Torr, George Allsop and John Lea 'do hereby own and acknowledge George Briddon to be an Inhabitant legally settled in the Parish of Brassington.' This declaration is witnessed by Robert Briddon and Thomas Scattergood and addressed by two JPs, Joseph Hayne and Robert Dale, to the churchwardens and overseers of Hopton. Hayne and Dale certify that Scattergood has sworn on oath that he saw the Brassington officials sign the document. With the Hopton overseers assured that Brassington undertook to take responsibility for him if he fell on hard times, Briddon was allowed to set up his forge there.

What happened to those who did fail can be seen in the removal certificates. Eight years before George Briddon was allowed to move from Brassington, Ralph Slack and his family were hauled back there.[5] They had tried to settle in Tibshelf and the Tibshelf parish officials had tried unsuccessfully to move them to Shirland in 1753.[6] In the following year the parish was successful in getting the family moved to what was, in fact, their home village. The removal order is addressed to the churchwardens and overseers of Tibshelf and Brassington:

...Ralph Slack, Martha his wife, Thomas and Ann their children are lately come into the said Liberty endeavouring to settle themselves there without giving Notice in Writing to the said church-wardens or Overseers of the Poor, or any of them, of the House of their Abode in the said Liberty... and... are likely to become chargeable to the said Liberty of Tibshelf. And whereas, it appeareth... that they were last legally settled in... Brassington. These are therefore in his Majesty's name, strictly to charge and command you... to remove and convey the said Ralph Slack his wife and children named above from... Tibshelf... to... Brassington and deliver them unto the Church-Wardens and Overseers of the Poor there...

Note that Ralph and his family were simply 'likely' to need relief. This was established by an examination before a JP – a miserable and degrading ordeal for a man hoping to make a living for his family in a new village. Ralph was twenty-six when he and his wife and young children were bundled into a cart by the Tibshelf constable and driven back in disgrace to their village.

What often happened to a pauper on his arrival home, or when he stayed overnight in another village on his way, was described by the magistrates at a Quarter Sessions at Derby on 13 April 1790 and published in *The Derby Mercury* on the 29th. The magistrates noted:

That it appeared to the Court, that there was wanting in most parishes in the said County, a proper Place of Reception for Vagrants and Paupers, immediately on their Arrival in their own Parish, with a Pass or Order of Removal; also for Vagrants and Paupers passed according to Lawe in the Way to their Place of Settlement, who were often in ill Health, and were, by being denied admittance to a Public House, sometimes put into Stables and Outhouses, where their Maladies encreased, their Lives were endangered, and Death often ensued.

Parishes were instructed to lodge the paupers at inns and bear the costs. If a pauper was in custody, on his way to his own parish, the allowance was to be 1s a night, and if he was too ill to continue his journey, the county would bear the extra expense. It was to be thirty years before Brassington had its 'place of reception', and meanwhile homeless paupers like Ralph Slack and his family were lodged at one of the village inns, perhaps the one which the overseers eventually bought as their house for the poor.

SCHOOLS

There was little schooling in the village for the boys and none for the girls. The curate, as we have seen, was expected to teach 'poor boys gratis' and the vicar's answers to the questions in the bishop's 'primary visitation' of 1772 included 'I instruct the Children in their Catechism & use my best Endeavour to fit & prepare them for receiving confirmation.' To a question enquiring about a free school in the village he replied, 'There is about 3 pounds a year that was left by Mr Dale a few years ago to teach children to read.' The vicar was slightly understating things. Thurstan Dale, one of Brassington's absentee landlords, had provided in his will in 1742 for an annual payment of £10, the proceeds from 4.75 acres of land. The money was to pay the salary of a schoolmaster. One of the earliest and perhaps the first to draw this pay was John Hardy. Hardy, a bachelor, died in 1758, leaving his house, household goods and 'clothes in one chest and two boxes' to his nephew. The pupils of a later schoolmaster, John Johnson, appreciated his work in the village enough to say so on his gravestone when he died in 1805: 'A few of his pupils in grateful acknowledgment have erected this stone.' Johnson was probably Hardy's immediate successor, as he witnessed a will as early as 1765. He was twenty-eight when

Hardy died and was schoolmaster long enough to have become a village institution. His schooling, however, reached only a few of the boys of the village. Half the villagers making wills in the second half of the century made their marks in place of signatures.

LAND TENURE

There were few living in the village who owned much land there. The eighteenth-century poll books, listing the voters in Parliamentary elections, demonstrate this. The qualification was to own freehold land to an annual value of £2 – the 'forty shilling freeholders'. In 1734 the Brassington list had thirteen names, only six of whom qualified by owning land in their own village – John Charlton, William Hodgkinson, Robert Lee, Thomas Marple, Anthony Walton and Robert Wilcock.[7] Thirty-four years later only two from a list of four had the required amount of freehold there – George Walton and the barmaster, Edward Ashton.[8]

The villagers rented land and houses from the copyholders of the old duchy-manor, and leased them from the owners of the former Shrewsbury Manor. The copyhold transfers are recorded in the manor-court books, which from 1700 consist entirely of these transactions, the court presided over by a steward employed jointly by the lords of the two manors. There are many examples of small copyhold estates, some held by outsiders, whose fields and cottages were occupied by villagers. Richard Gratton, a member of one of the old yeoman families, accumulated a copyhold farm and the copyhold of three cottages and arranged in the manor-court for their eventual transfer to his granddaughter Catherine Buckley and her intended husband Thomas Sleigh, a gentleman of Standow in Staffordshire. This was in 1701 and for the next forty-four years the court procedures were used successively for the inheritance, mortgage and sale of the property. On each occasion the villagers who lived in the cottages and farmed the fields are named. In 1731 they were William Beresford (one house, plus fields called Spellow and Banlong Furlong), Elizabeth Tomlinson (cottage), Henry Watson (cottage), Anthony Spencer (Catgreave, a croft and a barn), George Rippon (Middleton's Croft), John Scattergood (Suckstone Close), Mary Briddon (Shelbroad), Richard Buxton (Brackendoe), James Rowbotham (Cross Green) and George Chapman (Blackfurlong).

Through the century these villagers continued to rent their houses and fields from absentee landlords. Others joined the ranks of the copyholders. The court books make it clear that, although much of the village land was owned by a few larger landowners, nevertheless many of the villagers had by the eighteenth century acquired small acreages of their own, and that they used the machinery of the manor-court to pass it to their widows and children, to buy and sell among themselves, and to raise money by mortgaging it. This increase in some villagers' possessions is reflected in the number who left wills – seventy-four in the first half of the eighteenth century, compared to forty-two in the preceding fifty years, in spite of the departure of the richer families. The continuing prosperity of the lead industry was the reason.

The buying and selling, the wide spread of ownership, was in the land of the old duchy-manor. In the other manor, George Savile's, everything was different. Savile owned the freehold and leased the land and houses to twenty-five villagers, as his predecessors had done. In 1748[9], by which time it was held by Savile's heir, John Gilbert Cooper, it amounted to 469 acres. The holdings were small, with the exception of the forty-one acres held by William Allsop, seventy acres farmed by George Allsop and, by far the largest operation, 163 acres held by Job Marple, son of Thomas, and, like his father, landlord of the inn which had been the New Hall and was by then the Red Lion.[10] Much more typical were William Kirk's seven acres at Catgreave, with pasture on the common land for two cows, Richard Walker's Longlands and Reeve Holme (eight and a half acres) with three 'beast pastures', and Robert Buckley's Peaselands (two acres).

Until the nineteenth century this manor remained as one estate and all its barns; houses and fields were rented by villagers. The rest of the village farmland was the freehold for which the Lanes, Westernes and Buxtons had paid chief rent in the early seventeenth century. By the eighteenth century the owners of this land included Sir Philip Gell of Hopton, the nearest aristocrat to Brassington, and Thurston Dale of Ashbourne, the landowner whose will in 1742 provided the annual £10 for a schoolmaster's salary.

Taking the probate inventories as evidence, there were many more villagers with a few cows than there had been a hundred years earlier. The tithe document of 1667 had noted that the miners owned cows and grazed them on land rented from the farmers. The result of this development can be seen in the forty-one inventories which included cattle in the period 1701-1750 — there had been only twenty-seven during 1651-1700. The tithe document had also commented on the fall in sheep farming and there were in fact no large flocks in the inventories between 1701 and 1750. There were, however, very many more men owning a few sheep than there had been in the previous half-century — forty-one compared to fourteen. Typical of the miners of the late-seventeenth and early-eighteenth centuries was James Spencer. He died in 1713, leaving his copyhold house, barn and garden to his wife Mary, who was admitted to them by a jury of her neighbours at a manor-court meeting on 15 April 1714. Spencer's inventory includes three cows and a calf, forty-three sheep and £5 worth of hay and corn. However, in spite of owning his own house and livestock, he and his wife lived in the most spartan style — the inventory lists only a grate, tongs and pothooks at 1s, pewter and brass at 5s, and a pair of beds at 5s. Other wooden furniture, not described, was valued at 10s. There is no mention of rooms, but the Spencers obviously lived very poorly, in a small cottage.

GOODS AND CHATTELS

The eighteenth-century picture of a workaday village, a village without squire or gentlefolk, given by the overall inventory valuations is accentuated if the valuation figures for household goods alone are considered. Most of the Brassington men and women whose goods were listed and described for probate were living in a similar style to the miners described above. Apart from John Buxton's inventory in 1703, where his household goods were put at £169, there were fifty-four up to 1750. In only one of these were the household goods valued at more than £20 — John Allsop the baker had his valued at £21 14s in 1731. This one in fifty-four is a great contrast to the ten in thirty-six between 1651 and 1700, nine of which were in fact over £30 and one as high as £71 1s 3d (German Buxton in 1652).

The baker, John Allsop, was making a better living from his trade than the miners from theirs. He was living in a house with two downstairs living rooms and two bed chambers above, plus a cellar, brew house, malt chamber and, of course, a bakehouse. Here was a baker who brewed beer and perhaps sold it. He also farmed. At his death, Allsop was keeping three pigs, two horses '& their Tackle', thirteen cattle and eighty-nine sheep. Allsop's house had no luxury, but there was some comfort, with five beds, six old chairs, a seat and a squab — a wooden settle with, usually, an upholstered seat — and three little settles in the cellar. There were also table cloths and napkins valued at £1 — an indication of a certain social aspiration in the Allsop household.

The tradesmen, among the leading men in the village, were living in considerably less style than such yeomen families as the Lanes and Goodwins had enjoyed in the preceding century. Even so, they had more possessions around them than most of the eighteenth-century farmers. Another John Allsop, who died in 1728, had a set of furniture and equipment similar to the miners' and it is notable that, although his description in the probate document makes him a farmer, he had in fact fewer animals than the miner James Spencer. Some families seem to have come down in the world.

In the late seventeenth century one of the most substantial men of the village had been the innkeeper William Dodd – his inventory goods were valued at £200. His namesake, most probably his son, who died in 1748, lived a much more straightened life. He was described as a husbandman and had ten cattle and fifty-two sheep. He and his wife lived in a four-roomed house – two up and two down. The 'house' contained a dresser, tables and some pewter, the parlour, a bed and a table. One chamber had a bed, five chairs and another table and the second chamber contained two beds and two chests, with linen in one of the latter. It is difficult to see why the couple's only chairs should have been in one of the bedrooms, and there are no fire irons in the list. Like most of the inventories, this one leaves unanswered questions. It is clear, however, that the Dodd house had no luxuries. Another husbandman with the surname of one of the more prosperous of the seventeenth-century yeoman was Joseph Lane, who died in 1729. His widow inherited a set of fire irons valued at 2s 6d, pewter at 4s, brass at 2s 6d, a 'dozen trencher' at 6s and 'an old smoothing iron' priced at 6d.

COMMON LAND

To men with only a few acres of their own, and with only the limited common pasture of Upper and Sides, the 3,000 acres to the north, west and east of the village were vital. As we have seen, the people of Elton and Winster also had grazing rights on Brassington Moor, and eighteenth-century farm sales advertised in *The Derby Mercury* drew attention to them. For instance, an advert on 18 January 1787 for a farm at Elton included 'an unlimited right on Elton and Brassington Commons'. There had been quarrels over numbers of sheep pastured there since the Middle Ages. The villagers of the eighteenth century used the courts to settle their disputes. In 1720 Aston Hill, near Pike Hall, was disputed by Brassington and Elton and there survives a sketch map in the Wolley manuscripts[11] with a note which makes it clear that the villagers were still using 'ancient custom' to make their points. On the side of a track called Salter's Way, which is now the Pike Hall to Winster road, the sketch shows a small triangle, with the note 'a sheep fold, formerly a garden, the walls of wch Brasson people threw down in going Procession & shewd the Jury'. 'Going Procession', or beating the bounds, was a very ancient method of letting it be known that the village knew its boundaries and would make sure that they stayed safe. The sketch also marks 'a guide stoop set up by Brassington inhabitants' to show to the jury. This was at Pike Hall, the limit of Brassington Moor and the wasteland which the village had always regarded as its own.

The map shows Salter's Way forming the northern, fenced, boundary of Brassington Moor, with a north-west boundary separating Brassington common land from 'Ballidon Liberty'. Running parallel to this boundary is the road from Buxton to Derby, through Brassington village. A sketch map of 1723[12] shows the whole of Brassington Moor from Pike Hall to Griffe and Harborough, with boundary fences and gates – Ballington (Ballidon) Moor Gate, Pasture Gate and others unnamed. The Ballidon boundary and the Buxton to Derby road are shown on the east of Minninglow. The whole area is criss-crossed with routes, though it is impossible to distinguish tracks from footpaths – the routes to Ashbourne, Chesterfield, Buxton, Derby, Winster and Wirksworth are marked. The moor was not wholly unfenced. There are boundary fences to the Duke of Rutland's (sheep) Walk on the edge of Aldwark, Cheverell (Sacheverel) Walk, next to the Winster Moor boundary, and a third walk of which the name is illegible, on the edge of Aston Hill. The boundary of Aston Hill is marked on the 1723 map as being an 'old wall', and the walk adjoining it has a 'new boundary' marked to its north-east, adding another seventy-two acres to its original 208. The Duke of Rutland's Walk covers 238 acres and Cheverell Walk sixty-four. The rest of the moor is given as 1,368 acres, making the extent of the northern part of Brassington's 'wastes and moors' almost 2,000 acres.

ENCLOSING THE MOORS

We can be sure that the villagers' zeal to keep their own was matched by their wish to keep the wastes common and open to their sheep. They had opposed William Savile's enclosing in the previous century and, before Savile's time, had assured the duchy of Lancaster that there was no gain to be had in enclosing the wastes. However, the larger landowners, outsiders with the power conferred on them by the land they held, were still, in the eighteenth century, working hard to get the moors enclosed – their large holdings in the village would ensure that they had the lion's share when the moorland was split into individual holdings. This enclosure was not achieved until the nineteenth century, but there is evidence that there were continual attempts before then.

One such effort was made in 1730[13], and opposed because the rights of Brassington's neighbours were being ignored. The neighbouring villages were represented by Bache Thornhill, who wrote to Joseph Hayne, steward of the Brassington manors, on 6 February 1730 warning him of his 'warmest opposition' to any attempt to enclose the moors without an allocation of land for 'many other places besides Brassington'. Thornhill had been on his way to London when a letter from his father prompted him to write immediately to Hayne:

> This evening an express Messenger overtakes me on my way to London with directions from my Father to wait on you. The subject is this. The Inhabitants of Brassington have for some time been indeavouring by Act of Parliment to enclose their Common – those indeavours have for a good while been suspended, but are now, as my Father is informed, reassumed in order to obtain an Act this Sessions for that end.

These 'inhabitants', whilst not named, were the larger landowners in the village, Gell, Dale and others including Thornhill himself. There had certainly not been a village meeting, and a favourable vote, before a private bill was drafted to be presented to Parliament. The eighteenth and nineteenth-century enclosures were usually accomplished in the way that was being attempted in Brassington. A group of those who owned most of the land would rely on support from sympathetic MPs to get a bill through Parliament enabling the enclosure to take place. This was eventually achieved in Brassington but the 1730 effort foundered because the landowners were not united. The Thornhills protected the ancient rights of Brassington's neighbours to pasture on the moor. Bache Thornhill spelled it out:

> I give you the trouble of this, therefore, to assure you that if any steps are taken in this affair, without first coming to an equitable, fair agreement with the Inhabitants of Elton, Winster, etc, who have an indisputable right, we are ready and determined to give any such Bill the warmest opposition, & consequently from the nature of the thing, to defeat you.

Thornhill invited Hayne to write to his father to explain his plans and ended his letter with a cordial invitation – 'I shall rejoice to see you in Town, where you'll find me in Pump Court No. 6, pair of stairs, where be assured you'll meet an obliged humble Servant.'

Whether or not Hayne went down to London and along to No. 6 Pump Court, things had moved by November, when Bache Thornhill's father, John, wrote to Hayne.[14] He described a piece of ground which the Elton people demanded as the price of their agreement to allow the Brassington enclosure bill to go ahead:

> I therefore thought it most useful, in plain termes, to acquaint you, that the Inhabitants of this place are ready to permit Brassington to proceed provided that part of the

Common, which lies contiguous to Elton and which was proposed by the people hereof at your last meeting to be laid to that Place, be granted to them.

This sounds like Aston Hill, disputed by the villages ten years earlier. Thornhill refers to the ancient rights, with which the Elton people had always been happy, and claims that the Brassington people rarely used this part of the common, 'nor had they any occasion from the great tracts of ground lieing nearer to them.'

Thornhill reiterated that the Elton people would defend their rights, but gave it as his opinion that Brassington would be the gainer from the terms which they proposed, 'which are, in my opinion, highly favourable to Brassington'. Clearly Gell and his fellow landowners disagreed, and this attempt to enclose the moors failed. The miners and others of the 'poorer sort' were left free to graze their few sheep there, for the time being at least.

The end of Brassington's common land came with the successful Enclosure Act of 1803 and the award five years later. The 'allotments' made by the appointed commissioners showed that the pattern of landowning in the village had not changed during the eighteenth century. The common land was allotted in proportion to existing holdings of the 'ancient inclosed lands', and most of it went to a few outside landowners. The total area was 2,479 acres; 353 acres of this went to the Revd Philip Storey, the current 'improprietor', in lieu of his tithes.

The lords and ladies of the manor received one eighteenth of the land remaining after Storey's allotment. There were still two manors. The former Shrewsbury Manor had been sold in 1749 to Henry Coape, esquire, of Duffield, and had descended via Coape's grandson to his (the grandson's) cousin, Henry Sherbrooke, esquire, of Oxton, in Nottinghamshire and then to his son William. William Sherbrooke sold his manor in 1804 to Robert Lowe, who ended a very long story by breaking it up and selling it. In 1808, however, Lowe was still lord of his manor. The old duchy-manor lasted longer: it was left by William Newton to his two youngest daughters, Frances and Elizabeth, and by 1808 was held by Frances's widower, William Locker, another Frances and John Hayne, son of the third daughter of William Newton. The manor came back into the hands of village residents a hundred years after the death of its last resident lord, Richard Buxton, when Locker, by then the sole owner, sold it in 1824 to William Charlton and George Gregory. The eighteenths came to forty-four acres each and, with additional allotments, Locker and Hayne received 306 acres and Lowe 386. Two others were given more than 100 acres – Philip Gell (133) and Bache Thornhill (194). Five people, 3.5 per cent of the individual allotment holders, were granted 1,464 acres, or 59 per cent of the whole.

To continue the analysis, fifteen people received allotments amounting to ten or more acres and there were 116 with less than ten acres: forty-two of these last-named were given less than an acre, but the distribution meant, nevertheless, that many, perhaps most, of the village families, had some land. There is no agreement among historians about the effect of the enclosures on the small land-holders or the landless men. In Brassington we know that William Savile's seventeenth-century enclosing was resisted in court. The nineteenth-century Act, however, was not opposed and there is a record of only one individual protest.[15] Thomas Moore and his wife, from Bradbourne, were charged at the Quarter Sessions in October 1809 with 'pulling down a fence erected under the Brassington Inclosure Act'. They had presumably lost their right to pasture on Brassington territory. Curiously, while Moore was bound over for five years, his wife Ann was sent to the House of Correction at Wirksworth for two weeks and charged costs. We can only assume that she was more eloquent than her husband. In Brassington, for as long as their income included something from the mines, the small allotment holders were able to retain their few fields. Among these smallholders were family names which had figured in the manor's suite rolls and in barmasters' reckoning books for generations – Baker, Briddon, Crosby, Dakin, Harrison, Johnson, Lea, Lomas, Mellor, Oldfield, Roose, Slack, Smith, Spencer, Taylor, Torr, Walton, Watson, Wright.

Grazing rights for cattle on Upper and Sides pastures and the 'moors and wastes' had gone with land ownership in the manors and the award reveals how it had been distributed before the enclosure. Thirty-three allotments were given in lieu of beastgates on the pastures and twenty-one in lieu of common right, amounting to 671 acres for the beastgates and forty-one for the common right. Of the area given in return for the beastgates, 632 acres went to five people, including 170 to Locker and Hayne for seventy-five beastgates, 228 to Lowe for 126, and 129 acres to Bache Thornhill for 400 beastgates. Two village landowners, Thomas Allsop and Robert Millington, had eighteen and thirteen beastgates respectively. Most of the rest were held in twos and threes by members of the Brassington farming families – the Allsops, Charltons, Toplises and Waynes. Three each were held by the miners Francis and Richard Fearn, but they were the exceptions. The cows in most miners' inventories had clearly continued to be grazed on their own small plots, on rented land or on the pastures by virtue of rented beastgates, since there had been a trade in them. Strangely, while the holders of grazing rights for cattle were compensated for their loss, those who held similar rights for sheep on the moor received nothing.

TRAVELLING ON THE HIGHWAY

For the first half of the century the village's farmers, miners and publicans kept the advantages arising from its position on the Derby-Manchester road. They were increased in 1738 by a Turnpike Act which improved the road from Derby. Turnpiking was the eighteenth-century attempt to solve the ancient problem of maintaining a decent road network. Local trustees undertook to raise money by tolls and to use it to pay for regular maintenance. The 1738 Act provided for improvement to the 'dangerous, narrow and at times impassable road' between Shardlow, where the London to Manchester road crossed the Trent, and Brassington, where it stopped. The reason for the turnpike ending there was that the route over the upland to the north of the village, the limestone plateau, was dry enough not to need maintenance. Turnpiking was clearly not going to achieve Roman standards, and the road had not advanced beyond Brassington when Burdett published his map of Derbyshire in 1789. An alternative route to the north through Ashbourne was turnpiked by the Act of 1738, cutting Brassington's advantage. A further turnpike in 1758, linking Oakerthorpe and Ashbourne, crossing the old road at Turnditch, must have redirected much northbound traffic westward to join the Manchester road at Ashbourne. By 1777 the trustees, while advertising in *The Derby Mercury* for 5 December that the annual incomes of the Osmaston and Markeaton gates were £279 and £115 respectively, were proposing to remove the gate at Knockerdown to the south of the village. Clearly there was no longer enough traffic through Brassington to pay for the cost of the gate and its keeper. The role which the village had played since its founders had built their huts near the Roman Street was over.

Travellers had a choice of inns and alehouses, as in earlier centuries. Newspapers had stories mentioning the Wheatsheaf in 1757, the Red Lion, kept by John Prestwidge, in 1761, and the George in 1768. In 1759 the barmaster, Edward Ashton, advertised his inn in *The Derby Mercury* for letting. This was the New Inn, bought in 1754 from Job Marple. By 1777 the JPs of the Wirksworth wapentake were granting licences to three Brassington innkeepers[16], a number which had risen to five by the turn of the century. The reference to the Wheatsheaf in *The Derby Mercury* in 1757 was an advertisement by a new landlord which, in addition to offering 'good Accomodations, civil Usage, and the most grateful acknowledgements', reminds us that the eighteenth-century English knew their place. The inn's services were offered to 'Gentlemen, Ladyes, and others'. The 'good accomodations' at Job Marple's Red Lion, which had been named in a lease of 1746[17], are amply set out in Marple's inventory of 1755. There was good oak furniture in the 'great parlour or dining room' and in the eighteenth-century equivalent of the tap room. The dining room had decent crockery, including flowered china coffee cups, and was decorated with pictures and maps. There were six bedrooms, with ash-feather beds, and the kitchen and cellars were well stocked with food and drink.

BUILDINGS

The building of today's limestone village started in the seventeenth century and continued through the eighteenth. By the time Brassington was surveyed and mapped in connection with the enclosure of the moors at the end of the century the village looked much as it did throughout the 1800s, before the building of the new school on Town Street in 1872 and the council houses near it in the twentieth century. On the east side of Town Street, Rakehouse Farm was built.[18] A group of barns was added to Sycamore Farm, which itself dates from the previous century. The miner's cottage behind Wash Hills Farm was built during the eighteenth century, as was the house on the east side of Town Street called The Green and the one on the west next to Green Cottage. Brassington Hall, on the north side of Well Street, was built during the seventeenth century. Two more manor houses were added in the eighteenth, one on the south side of Church Street, opposite Ivybank, in 1774, and the other on the north side of West End, in 1793.

Another farmhouse built in the eighteenth century was Bucksleather House, whose ancient name was changed to Brookfield in the twentieth century. One of the village's two remaining pubs, the Miners Arms, was built in the 1700s. A full description of it was given in the manor-court book when it was sold in 1771:

Manor House, Church Street.

West End Manor.

Brookfield House, on Nether Lane (formerly Bucksleather, from a nearby field name).

...all that messuage house cottage or tenement in Brasson aforesaid with a barn and a stable thereto belonging. And also so much of a garden (adjoining the said house) as extends to the middle part of the middle window in the said house... and also all that messuage house cottage or tenement adjoining to the north-west end of the aforesaid house commonly called Palmer's House.

The Miners Arms, like other buildings mentioned here, is a listed building and although the list gives its date as late eighteenth century, Robert Wayne, who sold it in 1771, had bought Thomas Palmer's house in 1725. The pub is an amalgamation of a number of formerly separate buildings and at least one of its parts was clearly built early in this century. Other eighteenth-century additions to the village were Pleasant House, opposite the Miners Arms, and Church Gate Cottage, on the north side of Church Street. This house, like many more in the village, used to be two cottages. With the turnpike road came toll houses. There was one at Hipley which was demolished in the 1960s and one on the Aldwark Road which was for part of its life a cow shed.

On a more modest scale than neighbouring Wirksworth and Winster, Brassington was benefiting from the prosperity of the eighteenth-century lead industry. Most of the surviving cottages were built in that century by men who, poor as they were, had made a better living from prospecting than they would have made as farm labourers, and the large pubs were built to cater for them, as well as for travellers.

LAW AND ORDER

Like any other small place Brassington was fairly law-abiding. Through the centuries the villagers have committed affray, not murder and theft, not highway robbery. However, they made their share of work for the constable and magistrates. The Quarter Sessions which in 1790 instructed parishes to pay for lodging paupers at public houses also instructed them to establish a secure place, with bars at the windows, to house miscreants. Anxious to save public money, they suggested that a pauper might be housed there until it was needed for a criminal. The reasons for their concern were:

That for want of a proper Place of Confinement in Parishes, riotous, drunken Persons taken up by the Constable for being disorderly, were carried to an Alehouse, as the only proper Place of safe Custody; and Persons charged with Bastardy were frequently detained at Public Houses, and half the Parish under Pretence of guarding them, (the Warrants being frequently served on a Saturday Evening) were entertained and intoxicated for several Days at his Cost, nearly to his Ruin, and totally to the Destruction of Industry, Peace, and good Order in the Neighbourhood.

It is unlikely that Brassington escaped this saturnalia on the rates, and William Wilcock wrote on 9 April to the justices' clerk: 'You'll be pleased to Acquaint the Justices at the Sessions that we have repair'd a house for the Confinement of Vagrants into Good Order.'[19]

The villagers sentenced to spells in the House of Correction, which was built in 1791 on Wash Green at Wirksworth, had an uncomfortable time. An account of it written in 1805 describes two small courtyards, one for the men and one for the women, two dayrooms 11ft by 7ft and, attached to each dayroom, two 'sleeping cells', 7ft by 5ft 6in. The latter were ventilated and lit by an iron grating over each door, 18in by 15in, and a 6in-square aperture in the door. The authorities made no allowance to the gaoler for food, and the prisoners relied on his good will if they were not to starve. There was 'no water but what is fetched from a distance'.[20]

Hipley toll house, on the Haddon to Bentley Turnpike (now the B5056) near the junction with the road to Brassington.

The parish constable was expected to fit his law-enforcing duties into his normal round. Thomas Slack, the barmaster, would have been expected, during several terms of office as constable at the turn of the seventeenth century, to postpone his reckoning and viewing while he asked questions, posted reward notices or did whatever he thought was needed, when his neighbour's coe was burgled or a horse disappeared from its pasture. The system did not work of course, and at least one Brassington constable, William Wayne, described in the Quarter Sessions records as a labourer, was fined 5s for 'wilfully permitting James Rowbotham to Escape out of his Custody' in 1764.[21] *The Derby Mercury* carried many offers of reward from victims of theft for any information which would retrieve their losses.

Two guineas (42s) was offered by the Mayor of Derby in 1777 for the capture of three escaped prisoners. One of the three was Robert Allsop of Brassington, who had been working as a framework knitter in Derby – 'a tall thin Man, about 5 Feet 10 Inches high, wore his own slank (thin) Hair; and when he absconded, was dressed in a light-coloured coarse Cloth Coat, or a high-coloured Claret, with Waistcoat and Breeches of the same.' William Wilcock and the barmaster Edward Ashton offered only one guinea in 1778 for help in catching thieves who had broken into the 'coves' (coes) of Messrs Crosby and Rouse and George and Edward Fearn and stolen mining tools. Also missing from the Fearns' coe was a pair of 'linsey drawers' – the miners kept working clothes as well as tools in the huts they built close to their work. The coes, isolated on the moor, must have been easy targets.

For centuries, villagers in every part of the country had persisted in regarding rabbits and other small animals as fair game, for themselves as well as the farmers and landowners. Occasionally they were caught. At the Quarter Sessions on 15 January 1799, Daniel Fearn, tailor, and Robert Briddon, labourer, were convicted of 'using a certain dog called a Lurcher to kill and destroy the Game, not being qualified by Law'. The only ones qualified were those who had paid £3 13s 6d for a game licence.

One attempt at improving detection rates for real crime (poaching was not mentioned) was the setting up of an Association for the Prosecution of Felons. Many villages tried this at the end of the century. The Brassington one was formed in 1796, and reported in *The Derby Mercury* of 31 March. It met annually at Ralph Lomas's inn, charged an annual subscription of 5s, and set up a twelve-man committee 'for directing prosecutions and settling other business of this Association'. It had an initial membership of forty, a good indication of the prevalence of theft, as it included most of the farmers and shopkeepers and the leading men among the miners. The Association had a table of rewards for help from non-members, ranging from five guineas for murder, burglary and highway robbery to 5s for stealing fruit during daytime. It used *The Derby Mercury* for its reward notices.

Sometimes the victim offered his own reward as well as the Association's. When George Toplis's butcher's shop, for instance, was burgled during the night of the 21 January 1797, and salt, knives and other tools taken, Toplis added 'the sum of Five Guineas, over and above the reward which will be allowed... by the society for prosecuting felons in Brassington.'

Toplis's burglar was not caught. If he had been, he would undoubtedly have ended on the gallows at Derby. A later group of burglars in the village suffered a slighty less final punishment when they were caught and successfully prosecuted. The 'bloody code' of the eighteenth century prescribed the death penalty for over 200 offences. It was revised in the 1830s but had been applied increasingly less rigidly for some time before then, mainly because juries were prone to find a man not guilty to avoid sending him to his death. When John Allsop's butcher's shop was raided in 1831, and seven sides of bacon stolen, the ringleaders were packed off to the penal colony in Australia.[22] All the burglars were labourers from Kirk Ireton. Rowland Allsop, a labourer from Bradbourne, made the same journey in 1838 when he was caught stealing sheep.[23] Transportation often meant death on a disease-ridden slow boat to Australia but Allsop survived the journey. While his Brassington wife Hannah and their two small children moved in with her parents at the Royal Oak on Well Street, Allsop served his ten-year sentence, married, and started a new family at the other end of the world.

George Toplis was a committee member of the Association for the Prosecution of Felons. The crimes which he and his fellows were prosecuting were the ones carried out by poor and often hungry labourers. The men of the Association, or some of them, had their own failings. *The Derby Mercury* of 19 October 1797 reported the cases of George Toplis and two other Brassington butchers, a baker and three other shopkeepers in the village, all members of the Association. Each had been convicted at the Quarter Sessions of 'having in his dwelling-house certain weights, not being according to the standard in the Exchequer' – in short, cheating his customers.

FUN AND GAMES

The Derby Mercury had a column of local news which over the years included the occasional item about Brassington. Three of these pieces describe the village celebrating big events. For the coronation of George III on 22 October 1761, 'a large Subscription was raised by the Gentlemen, Farmers, Tradesmen and Miners of the said Town, who bought a fat Cow, which was roasted for the Publick'. It seems from this that the lead miners were a prosperous-enough element in the village to be called upon when money was needed. The same report reveals that the village band had a long history – 'a Band of Music, consisting of two Hautboys, and Bassoons, with First and Second Fiddles, by very good Hands, play'd before the People round the Town'. The villagers joined in the choruses, the bells rang all day and, loyalty confirmed by the good beef and, no doubt, the good ale, 'the Evening concluded with the loudest Acclamations of Joy and Loyalty'.

The village celebrated the centenary of the Glorious Revolution of 1688 with a similar feast – 'few Places surpassed the Village of Brassington'. The fat cow for this day was given by the 'principal Gentlemen', along with 'many Hogsheads of Ale'. In March of the following year, 1789, the village took the king's recovery from one of his periods of insanity for another feast day. There were bells ringing, a parade by the band and the church choir, who sang a new song specially composed for the occasion, a bonfire 'which contained about five Ton of coals', and 200 gallons of free beer. Not surprisingly 'the evening concluded with the greatest harmony'.

The Mercury advertised one savage old amusement which has long been forced underground – cockfighting. There are a number of notices of meetings at the house of Gervas Wood at Pikehall, including a match between 'the Gentlemen of Derbyshire and the Gentlemen of Yorkshire' on 16 and 17 April 1777. Twenty-five cocks were to be matched at two guineas a 'battle'. Two years later, for a match between Derbyshire and Staffordshire, the prize had doubled and there were to be three 'bye bye battles on the Saturday morning' to round off the sport.

REFERENCES

1 LJRO B/A3
2 LJRO B/v/5
3 DRO D166M BoxA147
4 DRO D258/45/18c
5 & 6 DRO Removal Order
7 MLSL Derbyshire poll book, 1734
8 DRO QS Derbyshire poll book, 1768
9 MLSL BM Add MS 6690 f1 7
10 DRO D161B/6/66
11 MLSL BM Add MS 6694 f251
12 MLSL BM Add MS 6687 f3
13 MLSL BM Add MS 6673 f220-221
14 MLSL BM Add MS 6673 f223-224
15 DRO QSz book 4
16 DRO QSLz
17 DRO D161B/6/66
18 DCC Listed Buildings
19 DRO Q/AB 108
20 Hackett, 1863
21 DRO QSz book 2
22 DRO QSz book 5
23 DRO QSz book 6

CHAPTER 9

CHAPELS, SCHOOLS AND HOUSE OF INDUSTRY

ENCLOSURE LANDSCAPE

Enclosure gave Brassington a new landscape. From west of Rainster Rocks to Carsington Pasture on the east and to the northernmost point of the parish at Grange Mill, the old moorland vanished, transformed into fields. These were bounded, not by hedgerows like those of the seventeenth-century enclosure in the south, but by limestone walls. These miles of drystone walls rapidly became as much a part of the scenery as the old hedgerows or the surface limestone from which the stones were quarried. The enclosure commissioners stipulated which of the new owners was responsible for each wall of each field and, according to John Farey, the walls, at a height of five feet two inches, cost 6s per rood (fifty-one and a half yards).[1] The stone came from quarries, listed in the award. Two of them were at the sides of Manystones Lane and the Wirksworth Road respectively.

The Enclosure Act of 1803 specified 'the Open Fields, Meadows and Pastures' as well as the wastes. The enclosure map identifies open fields bordering Ballidon Way and, at the south of the village, the Green. This field contained a large mere, the main water supply for the village farm animals, and the main well in the village, then known as the Green Well and earlier, for centuries, as the Coole Well. The Green was allotted to George Toplis and Farey describes how Toplis channelled water via land drains to 'some very complete watering places', though he does not say whether these were exclusively for Toplis's use or were also open to others. The well and mere remained accessible from the highway, and were in use until the twentieth century, when the mere was filled. The Green itself was used as a sports field until houses were built there near the end of the twentieth century.

It is apparent from the parish map, undated but probably drawn about 1820[2], that the large allotment holders wasted no time in splitting up their fields into more manageable ones. The 135 acres north of Manystones Lane which had been allotted to William and Frances Locker and Joseph Hayne, lords and lady of the duchy-manor, was fenced into thirty fields still there today. Thomas Allsop's thirty-eight-acre allotment between Manystones Lane and the Wirksworth Road became the four fields there today and his thirty-four acres south of the Wirksworth Road today's three fields. There are other examples of the reshaping of the commissioners' awards, but the result of their work, carried out between the Enclosure Act of 1803 and the Award of 1808 was a field pattern which has hardly been altered since then.

Also surviving, but in many cases in poor repair, are the barns and cow sheds built by the new allotment owners. Typically, these have stalls for half a dozen cattle at each end, facing into a central area called a 'bing' from which the cattle on each side were fed. Overhead was a storage loft for the fodder. Some of them have found themselves transformed in recent years into desirable country residences.

Most of the new fields were from the beginning used for pasture, including 'ley' or summer pasture, when fields were let at prices per head of cattle or sheep or per horse.

Typical enclosure walls and barn; in the bottom right of the picture is a pond, giving access to livestock in three fields.

The commissioners themselves raised money for the expenses of their work by letting Brassington Common in the summers of 1806 and 1807, before the allotments were made. This information is from Farey, who goes on to say that one of the new proprietors, with a 350-acre allotment, did the same the following year. This was probably the Revd Phillip Storey's allotment at the Tythe Farm, near Aldwark, which the parish map shows split into twenty-three fields.

At the beginning of the nineteenth century the village fields had not all been given over to meadow or pasture. Some were still being ploughed and cropped, and Farey describes how parts of the newly-enclosed rough ground of the wastes were prepared for sowing with oats and turnips. Most of the wasteland was covered by heather or moss. This was 'pared' by a tool called 'a paring-shovel or breast-plough, which a man shoves before him', gathered into heaps, and burnt. This back-breaking operation was followed by spreading lime, made on the spot. Some of the limekilns can still be found.

The enclosure commissioners also specified a list of roads, bounded by the new walls and to be maintained by the allotment holders with stone from the new quarries. Most of those designated as 'public carriage roads', to be 30ft wide, are still public roads. In or near the village were Pikehall Road, the continuation north of Town Street, Manystones Road, one of the two roads to the east, and Yearnstones Road, connecting Kingshill and Pikehall Road and now restricted to village traffic. One, the Wirksworth Road, was closed about twenty years ago and two others, the Hoe Grange Road and Galleylow Road, both running from Pikehall Road to Hoe Grange, were not built. There were also to be 'private carriage and drift roads', 20ft wide. Two of these were built and are still used – the Mere Road, which has since become Lot's Lane, and Sydes Pasture Road, or Wester Lane. Sydes Pasture Road ran to the allotment made to one of the bigger farmers, Robert Millington, and Millington was also to have a road north from his allotment to the Wirksworth Road. This was never built. Two others which never materialised were Slack Road, 'branching out of Yearnstones Road, north and north-west to Esther Slack's house', and Charlton's Road, 'from Ballidon Road over

George Oldfield's allotment to a gate in William Charlton's old enclosure'. Town End Road, described as running from Yearnstones Road to Pikehall Road, may be the lane now known as Bowling Green, which connects Kingshill and Dale End – Yearnstones Road reached that far in 1808.

ROADS

The responsibility for the upkeep of the roads, so long and so unsatisfactorily carried out by 'statutory labour' drafted in by the annually-elected parish surveyor, was transferred to a paid highway surveyor in 1835. This was under the terms of a Highway Act which provided for the setting up of highway districts and the levy of a highway rate to pay the surveyor and the labourers he employed. The surveyor's accounts for the early 1890s have survived[3] and show how the roads were being maintained then. The materials account lists the 'labourers employed in getting, carrying, or preparing materials', and the material used, which was broken stone in every case but one – the exception was gravel. The men were paid 1s 9d or 1s 10d for enough stone to pave a yard of road and usually supplied about thirty yards. The stone was brought to the roads by farmers paid 5s a day per horse used, and the finishing and maintenance was carried out by labourers paid 2s a day, a wage which was raised to 2s 6d in 1891. These men filled holes, levelled the stone, raked out ruts, scraped the roads, cleaned the sides and ditches, loaded dirt onto the farmers' carts and signed their names in the surveyor's book when he paid them at the end of their day. This expenditure of time, sweat and money produced roads which were far better than in the old, haphazard days, but which were still dusty in summer, muddy in winter, and always bumpy. Only the introduction of tarmacadam in the twentieth century cured these ills.

NEW GOVERNANCE

The Highways Act, the establishment of the Ashbourne Poor Law Union in 1845, the setting up of the Derbyshire police force in 1857, and the formation of the County Council in 1888 were parts of a process which, during the nineteenth century, removed from the parish the local powers which it had taken from the manor-court in the seventeenth and eighteenth centuries. At the end of the century the Local Government Act of 1894 provided for parish councils where there was a population of over 300. The new councils were mainly advisory but Brassington Parish Council's accounts in the 1890s[4] show that it soon began to use its powers to keep the village smart and clean, weeding and cleaning footpaths, carting away rubbish from the Green and emptying and repairing the parish 'watering'– presumably horse troughs. The councillors raised a little money by charging 1s a head for six stalls in the street on 2 September and 4s the next day, during Wakes Week.

THE EFFECT OF THE END OF MINING

The nineteenth century saw the end of lead mining in the lives of most of the villagers. It lingered on into the twentieth because the urge to go prospecting was still strong, but the production figures tell a story of decline. During the first half of the century the Brassington miners sold between 200 and 500 loads a year, less than their ancestors had sold in William Blackwall's days, and less than half the output during the eighteenth century. By the late 1870s the production had dropped below 100 loads a year. A rich vein had been struck at the Victoria mine in the late 1840s, and the annual output rose to more than 1,000 loads, with one year, 1848, showing a figure of over 2,000. These high figures were boosted by the strike at the Nickalum mine in the late 1850s and over 2,000 loads were again produced in 1861 and 1862, but the revival was brief and the industry had disappeared as a significant economic factor well before the end of the century.

The effect on the villagers of losing the income from their little mines is shown by comparing land and house ownership at a date in the earlier years of the century with those at its end. The enclosure of 1808 had included 116 allotments of under ten acres, and Glover's *History, Gazetteer and Directory* noted in 1829 that much of the land in Brassington was 'occupied by the owners' and that there were 'numerous small freeholds' in the village. The first document to give a precise picture of who owned what in Brassington was a Poor Rate and Land Tax Survey drawn up in 1835[5]. The parish then covered 4,023 acres. Of this, 64 per cent was owned by outsiders and 68 per cent was being farmed in units of fifty acres or more. However, in spite of this concentration of land ownership, there were nevertheless 160 personal owners. Of these, seventy-nine were living in the village and seventy of these were living in their own houses. Eighty-six householders were living in rented houses and nine of these owned other property. In 1835, then, although twenty-six of the villagers owned no more property than the house they lived in, and a further twenty-five owned less than five acres, Brassington was still a village in which almost half of the heads of household owned something. Things were quite different by the end of the century. A rate book of 1893[6] gives these figures: total acreage, 4,144; proportion owned by outsiders, 80 per cent; number of personal owners, 173; number of owners living in the village, forty-nine; number of villagers living in their own houses, twenty-nine. By 1893, 117 householders were living in rented houses, and of these twenty owned other property. The proportion of villagers living in houses which they owned had therefore dropped from a half to a fifth and of those owning land from a third to a fifth.

The decline in mining is reflected in the number of men described as miners in the census returns. The first return which gave every person's occupation was the one for 1851. At that date there were still forty-three miners, most of them probably not working in the traditional small mines but at the currently-prosperous Victoria mine. By the time of the 1881 census the number was sixteen and from the production figure for that year — seventy-six loads — it seems likely that these men were in fact earning their bread more from other jobs than lead mining.

FARMING

The census figures reflect another fundamental change in the lives of the villagers. There were seventy-six 'agricultural labourers' in 1851 and only thirty-six in 1881 — the consequence of the final disappearance of arable farming in Brassington. At the time of the enclosure of the wastes, as we have seen, the farmers were still growing oats and turnips. Both were fodder crops and John Farey also mentions the crops of potatoes grown by George Toplis as fodder for his bacon pigs. Brassington, according to Farey, was 'famed for its Bacon, about 400 Hogs being now killed there annually, and cured for sale.'

George Toplis, John Farey's chief informant in Brassington, was the biggest farmer in the village, and it can be assumed that his way of using his land would be the most profitable for the rest of the village's farmers. His main activity, apart from pig keeping, was dairy farming. Dairy farming remained the main business of the farmers throughout the century. When Clipshead Farm was sold in 1877, for instance, the catalogue spoke of 'the well known cheese making and dairying district of Brassington'.

As well as the oats, turnips and potatoes there were also fields of flax in 1813 and though in 1829 Glover characterised the farmland as '4,073 acres of good dairy land, chiefly meadow and pasture', there was still ploughland at that date. *The Derby Mercury* advertised the sale of '56 acres of arable, pasture and meadow land' in January 1830, farm auctions in 1831 included ploughs, harrows, a land roller and winnowing fans, and when the George & Dragon was sold in 1831, the property included 'copyhold arable' in its forty acres of land. Arable farming required more labour than cattle or sheep, and once the fields were grassed the only time when large numbers were needed was at the hay harvest, when the farmers paid a shilling for a day's work.

NEW JOBS

As the century progressed, although there were as many farmers in the village as in 1851, many more of them were farming rented land. There were fewer villagers with their own houses or fields, fewer with their own mines and fewer working in other men's mines or on other men's farms. There were two new ways of earning a living – maintaining the track, bridges, viaducts and embankments of the Cromford and High Peak Railway and running the 'wharf' at Longcliffe, and quarrying. The railway's need for stone added to the demand already created by the building of the miles of enclosure walls. The first section of the railway, from Cromford to Hurdlow, was opened in 1831, though it had reached 'Brassington Moor Wharf', at Longcliffe, by 1830, when it was reported in *The Derby Mercury* that a Mr Frost, rope-maker of Wirksworth, had rented it for the sale of coal. The line passed through two and a half miles of land in Brassington and at the survey of 1835 the company owned over twelve acres, with a house, a counter house and a 'cole yard'. The 1851 census recorded only a 'wharfinger' or stationmaster at Longcliffe, employed by the railway, but by 1881 the number had risen to ten. These men, who included the stationmaster, platelayers and labourers, were full-time railway employees. There were many more who worked on the construction and maintenance of the walls, bridges and viaducts for limited periods.

Quarrying became important to the village in the second half of the century. The quarries listed in the Enclosure Award were not included in the 1835 survey and there were no quarrymen in the 1851 census return. By 1881, however, there were eleven quarrymen living in the village and the 1893 rate book lists a quarry at Manystones rented by Joseph Banks, sandpits at Roundlow and Wester and a 'sandworks' at Longcliffe, owned by James White and Co. This firm also had a brick kiln at Longcliffe, using the sand from its own pit and from those at Roundlow and Wester. These jobs, and the road-making described earlier, replaced part of the work which had for centuries kept the men and women of the village busy in the mines and on the farms, but it is plain that by the end of the century there was less work to be done. The 1881 census shows only eleven people unemployed but, significantly, there were thirty-nine fewer living in the village than there had been in 1851 – 690 compared to 729. Brassington had started its long decline as a working village.

Many left the village and those who stayed struggled to find work. An example is Thomas Slack, last of a line of miner/farmers, living in a copyhold cottage on Hillside and farming about four acres of copyhold land. His mining ventures failed during his last years, and when he died in 1850, without leaving a will, his widow was unable to pay the court fees needed to transfer his fields. His son regained the cottage a few years later, paying the court fees of £7 3s 10d with a down payment of £4 and the rest over five years. He and his mother raised this money mainly by selling his father's mines. He worked at a variety of jobs in and around the village and ended his life in 1909 farming a small-holding at Longcliffe. With the exception of his youngest son, all his children had left the village by then, and the son who stayed, working all his life at Longcliffe station, was the only one of his name in the village. There were many families who broke a long connection with the village and some attempted new starts across the sea.

Members of the Fearn and Hallam families settled in Albion, Edwards County, Illinois and their descendants are proud of their origins in Brassington. Not all were successful. Another of the Fearn families, settled to farming in Massachusetts, was wiped out by successive tragedies. One member of the family was murdered for his money as he rode home from market, and the rest were killed in a flood. More typical than this drama was a move within Derbyshire or to a neighbouring county as the chances of making a living in Brassington grew fewer.

SHOPS

The decline was slow, however, and for the whole of the nineteenth century and for much of the twentieth, there remained the shops, shoemakers, blacksmiths, joiners, butchers, tailors, stonemasons, dressmakers, cattle dealers, carriers, preachers and teachers who made it a lively and, to a large extent, self-sufficient community. The villagers had five resident coal merchants, a maltster, a straw-bonnet maker and two policemen. There are no bakers in the 1851 census return, but there had been earlier in the century: *The Derby Mercury* reported the accidental death of 'Mr Swindell, baker, of Brassington' in 1830, whose cart overturned at Pikehall while he was on his way from selling bread in Buxton, and, 'Mr S. was found early on Sunday morning, inside the cart, a corpse'. The account books kept by two of the village's cordwainers (shoemakers) have survived.[7] These were William Walker and his son, also William. The elder William's book starts in 1835. The Walkers had a large permanent clientele, coming regularly for a new pair of boots (about 13s for the men and 8s for the women), and for repairs. When the second William Walker died and the family business ended, there was left a great accumulation of their customers' 'lasts' or wooden patterns. Among them was a pair smaller than the smallest pair a child would wear, the pattern for the fairy shoes which the lead miners took underground with them, still half-believing that there were fairies there to wear them. The Walkers also made and repaired harnesses, including in 1859 a 'Lether to Windering fan' – there was still grain to be winnowed at that late date. There were five or six pubs for most of the century and, at least as good an indication of vigour among the villagers, there were three chapels.

CHURCH AND CHAPEL

At the time of the Primary Visitation of 1772, when the vicar answered questions posed by his bishop, he was able to assure him that 'there is neither Quaker, Presbyterian, Independent, Anabaptist, Methodist or Moravian in the Parish'. As there were also no 'papists' the vicar had a congregation which included the whole population of the village. They may not have been enthusiastic, but if they attended any service at all it was still the one held in the old church. The vicar reported that 'about 30' received the Sacrament. He held one service each Sunday and preached one sermon. This religious calm was almost over. John Fearn, a retired grocer celebrating his golden wedding in 1909, reminisced to the *Ashbourne News*[8] that his grandfather had been the first to open his house for Nonconformist preaching in Brassington. This was in 1795, and the preachers were probably from the Congregational church – Fearn was later a member.

One of the most vigorous elements in the village was the Primitive Methodist church. An obituary in the *Primitive Methodist Magazine*[9] for one of the early leaders of the Brassington church, James Handley Fearn, who was a stonemason, describes the beginnings. Fearn's parents 'like thousands more, though but seldom attending the parish church, still regarded themselves as strict church people'. However, 'their son was permitted to grow up in the ways of sin', which extended to disrupting Methodist meetings. In 1819 Fearn and others ('wicked companions'), gatecrashed a service intending to ridicule the Methodists. Instead, Fearn found himself converted. He joined the Church, much against the wishes of his parents – 'they could not bear the thought of their son being connected with a sect which was then everywhere spoken against'. Fearn became a popular local preacher, speaking 'with much pungency and power', and 'neither long journeys, rough roads, stormy weather, dark nights, nor small congregations, could deter him from punctually attending his appointments'. At a time when he was working ten miles from Brassington he would come home late on Saturday night and then walk ten or twelve miles to a Sunday service. The obituary describes him wading through the flood waters of the Derwent to reach one congregation.

The former Primitive Methodist chapel, on Hillside Lane.

Primitive Methodist preachers had a fiery enthusiasm which won over the working people, and the organisation and activities of the 'societies' which they set up provided the villagers with purpose, keen spiritual excitement and regular, lively social occasions. There were at first only a few at Brassington brave enough to defy the mockery of their neighbours by joining in the prayer meetings and love feasts, at both of which they would be expected to proclaim their faith and pray in public. Their designated meeting place at the time of the 'Return of Nonconformist Places of Worship' in 1829 was Rebecca Titterton's house in Maddock Lake, and the society had fifteen members.[10] By 1833 the Brassington society was part of the Winster 'circuit', and had eighteen members, still meeting in the open or at each others' houses.[11] In the following year they bought seventy square yards of ground for £3 from one of their members, Henry Taylor, and built a chapel.[12] The obituary of one of the chapel trustees, Richard Kirk, gives the atmosphere which the new religion had created and how the chapel building was achieved.

Kirk's house 'was the house of the preachers while he lived', 'he generally sung with tears streaming down his face, and his soul rejoicing' and he 'advanced the purchase money for the land, worked occasionally with his hands, and used every means in his power to forward it'. This homemade chapel was built on Hillside Lane. It had pews with doors, and the society made a little money for its upkeep by renting them to its members. By 1836 the number of members had risen to forty.

The importance of the church to the villagers by the middle of the century can be seen in the report of a chapel meeting held at Longcliffe in 1862. There were more than 200 there, about 175 of whom had tea in a tent decorated with flowers. There were addresses, prayers and singing and 'the young people enjoyed themselves'. The meeting was held on an evening in August, replacing the weekly united meeting of the three chapels which then flourished in the village. The others were the Congregational, in whose magazine, the *Independent*[13], the Longcliffe meeting was reported, and the Wesleyan Methodists.

The former Congregational chapel, at Dale End.

The Congregational chapel was built in 1846, in a much more ecclesiastical style than the Primitives'. Like the Primitive Methodist chapel, it was financed largely by its members. They installed a harmonium in 1853, and stained-glass windows in 1858.[14] The church was particularly active during the pastorship of the Revd J. Bartram, running a Sunday school and educational classes and keeping a record of its activities and of the births, marriages and deaths of its members.[15] Bartram gave an address to a meeting held in the chapel in 1860 to celebrate the success of the evening classes. The meeting was attended by 170 supporters from all sections of the village people. Bartram expressed a very modern-sounding philosophy, stressing the importance of freedom of conscience and saying that neither in the evening classes nor in the day school supported by the Church, was there any restriction on members of other churches attending. He was seconded by others, including a twenty-year-old village joiner and wheelright who was later to become a member of the Board at the school established in 1872. His speech was reported:

> Mr Joshua Taylor apologised for his working dress. He came only expecting to be a hearer, but the chairman saying 'All honour to the fustian jackets and hard hands of Old England', which sentiment was heartily responded to, he consented to go on, and, in a neat speech, advised the young men and youths to spend their leisure evenings at the classes. He thought it would be more creditable and profitable than lounging in the streets with 'dirty short pipes in their mouths'.

Unsurprisingly, Taylor was a Primitive Methodist local preacher.

The Wesleyans, whose thirteen members had been meeting at Robert Tomlinson's house at the time of the 1829 return, had built their chapel in 1852, with the financial help of John Smedley, owner of a textile mill at Lea. Smedley, in addition to founding the hydros at Matlock and building Riber Castle, conducted revivalist services in a marquee with which he toured the Matlock area.[16]

The former Wesleyan Methodist chapel.

While the three chapels were creating new congregations among the working people of the village, the old church too was showing a renewal of vigour. A new vicarage was built in 1857[17], with £400 of its costs met by subscription, and in 1866 Brassington finally became a parish, independent of Bradbourne.[18]

The parishioners had apparently been at loggerheads with succession of vicars. The evidence for conflict between vicar and parishioners is a letter of extreme bitterness written to the bishop in December 1876 by J.B. Littler, who had been vicar from 1865 to 1871.[19] The occasion of the letter was the death in November of Henry Alexander Campbell after a spell of almost two years as vicar. Littler wrote 'in the interests of the vicars of Brassington', the last five of whom he said had been 'maligned, scandalised, and persecuted by the leading parishioners', in Campbell's case to the death. Each of the others had left the parish 'after a few years' residence, that is, as soon as they could'. Littler himself had clearly had a bad six years in Brassington. However, he had an even shorter stay in his next parish, Llantrissant in Monmouth, as by the time he wrote his angry letter he was the vicar of Hayten. Perhaps his Welsh parishioners had proved to be as scandal-loving as he had found the farmers and shopkeepers at Brassington.

In 1879 the parishioners embarked on a two-year renovation of the church. They raised £2,000, according to *Kelly's Directory* of 1895, to repair and enlarge it. This was a lot of money for a small village to raise, but it may have been the fact that they could not raise the even larger sum which would have been needed to build an entirely new church, as was done at nearby Parwich, which saved the old one from destruction. Replacing dilapidated Norman churches with brand-new ones was popular at the time. Repair was overdue at Brassington – a photograph taken before the work started shows cracks in the south wall. An eighteenth-century gallery was removed from the west end, and on the north the church gained a second aisle, an organ chamber and a vestry. The flat roof was replaced by a pitched slate roof, and the church was refloored, in part by putting memorial wall plaques to use as paving stones. The Willcock family memorials suffered in this way, and it was presumably during the same operation that Ralph Marple's 1695 gravestone became part of the steps to the porch.

The church after the restoration, with an extended chancel containing a rose window.

The church before restoration and enlargement, with a crack in its south wall.

The Norman style of the south aisle contrasts with the Ashford marble pillars in the new north aisle in the background.

Ashford marble was chosen for the circular pillars supporting the arches between the nave and the new north aisle – local stone but a strong contrast with the limestone Norman pillars on the south. The chancel was lengthened and the east window replaced by a rose window given by the builders in memory of an earlier member of the family, William Knowles, who had died in 1859. A heating system was installed which consisted of a furnace under the central aisle with flues to the more remote parts of the building, and the church was lit by oil lamps.

SCHOOLS

The sober, serious, moral side of English life is the side which best characterises the nineteenth century, and Brassington's church building was one of the ways in which it was expressed in the village. Another, and one whose effects have proved permanent, was the extension of education to all the villagers. There had been schooling for a few of the boys at least since Anthony Kempe was selling horn books at the beginning of the seventeenth century. 'School' was probably a room in the schoolmaster's house for most of the time, though a sale advertised in *The Derby Mercury* in 1831 included 'a new building, now used as a schoolroom, but capable of conversion to a dwelling'. This was on Town Street. John Fearn, born in 1826, reported that he was educated first at a dame school, perhaps held in the building on Town Street, and later at a 'larger institution'. What he was taught can be seen from a book presented to an earlier member of the family, Francis, in 1789.[20] Francis Fearn was a lead miner/farmer and the book contains several pages of his accounts, including income and expenditure at the Great Rake mine in the 1850s. It started life, however, as an exercise book. There are pages of writing practice, mainly practice alphabets but also including two attempts at the line 'Beauty likes [sic] princes are subject to mortality', an improving sentiment but not likely to

The 1832 school, built by public subscription at West End.

improve Francis's grammar. Most of the exercises are calculations. The 'sums' are of land areas, time, money, weights and measures. There are apothecaries' weights, wool and cloth weights, ale, beer and wine measures and problems for Francis to work out: 'a person was born in 1758 how old is he this present year 1804' or 'A lent B £100 but has paid him back £47 14s 7d. What does he yet owe him?' and others, all successfully completed. This very practical syllabus is probably typical.

John Fearn's 'larger institution' was probably the school built in 1832.[21] This school, like the chapels, was a village initiative. The trustees of Thurstan Dale's charity decided that the village needed a school. They bought a piece of ground called Mark's Croft in the west end of the village. The seller was a village lead miner called Christopher Slack and they paid him £5 for it. The purchase price and the cost of building the school (£138) was raised by subscriptions from 170 of the villagers. The first headmaster was Christopher Slack's son Daniel[22], assisted by his wife Hannah, daughter of a former schoolmaster, James Swindells, and the initiative for the school may in fact have come from Daniel Slack. It is easy to imagine the young school teacher (he was born in 1807) complaining to the trustees of Dale's charity how difficult it was for him to do the job they were paying him for without a decent building, and that his father had a piece of ground which he was willing to sell to them.

By 1846, when the National Society conducted an enquiry into the state of education[23], the schoolmaster, a new one as Daniel Slack had died in 1845, was teaching eleven girls and forty boys in the one-room school. His annual salary was £29 11s, and the other costs of the school amounted to £34 17s. These were met by a combination of £25 from the National Society, the Dale charity's £12 and contributions from the parents. The Society commented that further funds were needed to cover the master's salary and to build a schoolhouse. Two years later came a report from the 'Inspector of Workhouse Schools' on the children in Brassington workhouse[24], of which more later, and on their schoolmistress. The children were not advanced but 'the proportion of Children who can read fairly is considerable'. The inspector made recommendations for the new workhouse in Ashbourne, then almost ready for occupation, and we can assume that they give an accurate picture of the syllabus and equipment considered appropriate not only for workhouse schools but for others too. The recommended books were an arithmetic book, a three-volume set of reading books and *Sullivan's Geography*. The inspector recommended parallel desks, a blackboard and easel, a large map of the world and a map of England. The three Rs and geography were clearly the school fare in the middle years of the century.

Religious rivalry, particularly between the Nonconformists and Anglicans, was a feature of nineteenth-century education and the Revd Bartram's open-mindedness was not always shared by the schoolmistress. In 1865 the current incumbent left her job and was reported to have told the children to leave the school, which would soon close, and to leave their various Sunday schools and join the one run by the parish church. The committee appointed Miss M.A. Gould in her place and the school reopened with eighteen children.[25]

The next step was taken after the passing of the Education Act in 1870. This Act, the first of a series which by 1891 had established free and compulsory elementary education, provided that elementary schools, controlled by local School Boards, be set up where school provision was inadequate. It was inadequate in Brassington, but the inspector supported plans for a new school then being formulated by the trustees and, for the time being, self-help continued and no School Board was set up.[26]

The building and management of the new school was in the hands of a fourteen-man committee. Building work went on throughout 1871, welcome work for local masons, labourers and carters, and welcome trade for the landlord of the Miners Arms.[27] He presented a bill to the committee in 1872 for ale which he had supplied to a list of workmen and waggoners since April 1871, starting with Henry Watson and John Yates, 'finishing the wall'. The committee seems to have sanctioned a daily intake of two pints of beer, at 3d a pint; when 'Mr Allsop's man', carting grit on 5 June 1871, drank three pints, the landlord noted 'if objected to he will pay the extra'. For the ceremonial opening of the new school in April, the landlord supplied three gallons of ale and 4s worth of spirits and, at a children's fête in the same month, five gallons of ale for the band. The village seems to have welcomed its new school.

The headmaster, Daniel Taylor, a thirty-three-year-old graduate of Chester Training College, came quickly to the conclusion that the village's efforts at educating its children had had little success. The school opened on 13 May and in his first entry in the school log, written four days later, the headmaster judged that 'the children are in a very backward state'.[28] There were forty-eight of them during the early days. On 5 July a further five 'scholars' were added to the register, 'supported by Dale's Charity £7 15s per annum'. The decision to remain independent of the state meant that the school depended partly on charity and partly on parents' contributions – 'school pence'. It struggled. A third entry on the first page of Taylor's log described another of the problems which bedevilled the first years of the school: '3rd August Reopened School this week – very thin attendance indeed owing to many of the children being away haymaking'. The villagers' attachment to education did not extend to leaving hay in the field when there were children available to get it into the stack. On 2 September there were 'many children absent on account of the corn harvest' and an even stronger indication that the parents had a pretty weak opinion of the school's importance is the log entry of 18 March 1873: 'Many children come late to school in consequence of having to go on errands; some even after the attendance has been taken.'

The results of poor attendance and lack of discipline were described in a succession of inspectors' reports. In 1878 the headmaster expressed his frustration in the log: 'The education act has not made much difference in this village as the authorities are

The 1872 school.

very lax in putting it into operation.' The authorities who were causing his frustration were the trustees, and in 1884 they attracted some of the fire in another bad inspection report. Order, reading, handwriting, spelling, arithmetic were all pronounced 'bad' and the inspector noted that there was hardly any instruction for infants. The managers were blamed for a poor supply of reading books and for inadequate heating. They had also appointed an insufficiently qualified teacher and 'the arrangement by which the Managers pay the Teacher's salary appears open to objection.'

The school's main problem was money – the lack of it. There was a grant of only about £60 from the Local Government Board to supplement the quite inadequate local funds and in fact the committee would have opted for a School Board in 1875 if it had not been for the persuasive opposition of the vicar, Alexander Henry Campbell, anxious to maintain the church's influence over the education of the village children. Perhaps this issue lay behind Revd Littler's accusations that Campbell had been persecuted by his parishioners. By 1894 the committee, unable to raise money for repairs to the building, was ready to hand over to the Board of Education, set up in 1889. This was done in the following year, and a new School Board appointed. In 1896 the new board, in fact largely the same men who had composed the old committee, agreed to spend £538 on repairs and improvements and appointed William Garner and his wife as headmaster and assistant mistress respectively, at a joint salary of £120 per annum.[29]

The new headmaster had advantages over his predecessors. The school was now properly funded – repairs were done promptly and indeed a new room was built in 1897. Schooling was free and compulsory – there was an end to the wrangling over paying 'school pence' and an attendance officer (John Fearn, paid £5 per annum) chased up missing pupils. Garner was an imaginative and enterprising teacher. The log describes his method of teaching the decimal basics of arithmetic with bundles of ten sticks. By 1899 he had succeeded in getting one of his pupils through the 'scholarship' examination and admission to Ashbourne Grammar School. He held evening classes in first aid for men and dressmaking and cookery for women. Among his initiatives were Christmas pantomimes, a string band and old scholars' reunion parties. By the end of the century the school, staffed by Mr and Mrs Garner and by Alice Fearn and Sylvia Knifton at quarterly salaries of £7 1s, was a lively and effective element of village life.

SOCIETIES

Also vital in the lives of many of the villagers were the friendly societies. They were flourishing at the beginning of the century and had probably been there in the eighteenth. They combined insurance against illness and injury with an agreeable element of pomp and circumstance. John Farey caught the flavour:

> I happened, on a Thursday in May 1808, to be at Brassington, on the anniversary of the Men's Friendly Society there, of 120 members, who had a standing stock of 300L out at interest: the Members, all clean and in their best clothes, assembled early in the fore-noon to transact their business in their Meeting-room, and being there marshalled, each with a tall white wand tipped with blue, walked in procession through the Town, preceded by a band of music and a large handsome blue silk flag, on which was painted the appropriate story of the Good Samaritan (the gift of some patriotic Lady of the district, I believe), to the Church, where all the most respectable Inhabitants of the Town and neighbourhood attended, and heard Service and a highly appropriate Sermon; from which, through streets crowded by approving spectators, the Members returned, in like order, to their Room, to dinner.[30]

The meeting room was probably the Miners Arms. The landlord there, Thomas Slack, included entries for payments to 'the Club' among the debts run up by his customers[31]. Later in the century a successor to the Men's Friendly Society, the Oddfellows, also met in this pub, where Thomas Slack's grandson, Joshua, kept the membership and account books.[32] The Oddfellows, a national organisation, were founded in 1810 and the Brassington club at the end of the century was the Charity Lodge, No. 22, of the Derby Midland United Order of Oddfellows. The contributions book, starting in 1887, records monthly payments by members of 1s 8d and benefit payments to widows of 6d per month. Like the club described by Farey, the Oddfellows held anniversary marches through the village, carrying a very large silk banner, the sort needing six men to hold it upright on a windy day. There is an account in *The Derby Mercury* of 23 August 1862 of an Oddfellows anniversary which sounds remarkably like Farey's. The members gathered at the George & Dragon on this occasion, marched to the church, heard the sermon and marched back to their 'lodgehouse', where they had dinner and speeches. Afterwards members of the band, the Wirksworth Volunteer Rifle Corps band, sang 'excellent glees', and then there was more marching and music in the streets 'of the picturesque village of Brassington'.

PUBS

Church and chapel, school and friendly society: these were the formal ways in which the villagers organised themselves. Among the informal institutions which made the village tick were the pubs. During the nineteenth century these were the George & Dragon, now Dragon House, the Miners Arms, the Gate, the Royal Oak, on Well Street, a new Red Lion, at the bottom of Kingshill, which is now Red Lion House, and the Thorn Tree, on Kingshill, now become Thorntree House. For a detailed picture of the role of one of the village's pubs in the nineteenth century we have the Miners Arms accounts book described earlier.[33]

School photograph from around 1900 with Mr Garner at the centre with his scholars.

These accounts – the landlord's 'slate' – make it clear that one vital function carried out by a village pub was in supplying credit at a time when there was a shortage of currency. The account holders were farmers, landowners, a churchwarden, a surveyor, as well as the miners. The credit was often over long periods – years, in some cases. An example is the account kept by the surveyor, George Nuttall. Headed 'Mr Nuttall & his man', the account runs from October 1824 until October 1827, reaching the sum of £6 4s 6d. A much larger sum, £28 12s 7d, was owed by the miner William Fearn during the period 1808-1811, and there are many others.

The Miners Arms supplied food, 'Eating for 9 at 1 each 0-9-0' to the 'proprietors of Wayne's Dream', for instance, in 1818. George Nuttall (and his man) stayed at the pub during 1824, 1825 and 1826 and the accounts give us the costs: 'To one Weeks' Board for Man 0-9-0', 'Breckfast of Milk to Mr Nuttall 0-0-4', 'two Dayes Eating for Man 0-3-0', 'Diner for Mr Nuttall & Alsop', and many more, running up a bill which was not paid until 1827.

Drink, food and mining are the subjects of the largest number of entries in these accounts, but the landlord of the Miners Arms and, presumably, the other pubs too, helped finance many other village activities. There was the house rent of £1 10s on the account of Benjamin Alsop junior, of Carsington, in 1819, five 'poor books' at £1 13s 3d each for the overseer William Charlton in 1818, '20 pecks of Tatoes at 8d 0-11-8' (William Greatorex, 1816), and 'one years Poor Rates for House 0-8-9' (William Green, 1819). There are debts for meat, coal, sheep, pigs, hay, for the food and drink supplied by a village grocer to a funeral, and for ale supplied by the churchwardens to the ringers on Christmas Day, 1815. There are large cash loans and interest payments ('Lent in cash 20-0-0' and 'Interest up to LD [Lady Day] 1828 0-10-0' on Jacob Johnson's account, for instance).

THE WORKHOUSE

For a period of twenty-eight years the village had another institution, quite different from the chapels, schools and pubs. This was the House of Industry which the magistrates had recommended in 1790 as a place where vagrants and paupers could be housed. A Poor Law Act of 1782, called Gilbert's Act, had ruled that paupers should not be sent to workhouses more than ten miles from their own parish, a ruling which encouraged parishes to form unions to administer the Poor Laws. Brassington, and many other parishes in the county, had been sending their paupers to Ashover workhouse and paying Ashover a subscription for the service[34], but in January 1820 a union based on Brassington, the Brassington Incorporation, bought the old Red Lion from James Swindell for £195, and converted it to a workhouse.[35]

The workhouse seems to have been occupied very soon after its purchase by the overseers, as one of them, Thomas Slack of the Miners Arms, recorded 'Ale to Poor House 5/-' in his accounts on 12 June. Expenditure on poor relief had risen sharply in the early years of the century. A return in the Fitzherbert collection in the Derbyshire Record Office shows a poor rate of £131 in 1786[36], while during the years 1801 to 1817 covered by a return in the Quarter Sessions papers, the lowest figure was £227 and the highest £616.[37] This was a period of high prices and low wages. Mining was in decline and most of the pasture on the moors had been given to the landowners by enclosure. During the wars with Napoleon the overseers were also paying for the families of the few men who had joined the army – these payments rose to £13 in 1814. After the wars' end farming entered a period of low prices and depression. It must have seemed to the parish that some of the £700 they were raising from the rates could be spent on the workhouse, thus cutting some of the outdoor relief.

When the Brassington Poor Law Incorporation, or Union, was wound up in 1844, and its responsibilities transferred to a new Ashbourne Incorporation, a *Derby and Chesterfield Reporter* article revealed ramshackle and corrupt proceedings in some of the parishes, the chief culprits being the overseers. A witness testified that the 'treasurer and clerk' was William Slack, of Brassington who, he said, was a schoolmaster and not a guardian. Either the witness or the reporter seems to have confused William, son of the barmaster, who was a farmer and an overseer, with Daniel, who was certainly the schoolmaster and not a guardian – at that time the parish overseers were responsible for collecting the money for poor relief while the Board of Guardians were responsible for allocating it, a division of duties which was apparently not being observed in the Brassington Incorporation, where the overseers were dispensing the poor money.

The nineteenth-century workhouses were grim places, appropriately nicknamed Bastilles. Their unfortunate inmates were at the mercy of the 'governor' and in 1830 a song attacking the corruption and immorality of the Brassington governor and his wife found its way into print:

> Come all you beggars far and near give ear unto my song
> I've something to relate to you, which shall not keep you long
> It's concerning Mr Sheeplouse, that man of mighty fame
> Has pinch'd the poor at the Bastille and thought it no great shame.

There are six verses and a chorus, singable to the tune of 'the Linconshire Poacher'. The governor's wife becomes Mrs Clambeggar ('clam' = starve) and she and her husband are accused of starving the paupers and stealing the money for their food, while Mr Sheeplouse is accused of fathering the baby of one of the inmates. The song sheet, now in Derby public library, is anonymous but the surviving copy has a handwritten note dating it to 1830 and ascribing it to 'Robert alias Robin Slack, sculptor, schoolmaster, fidler, dancing master etc'. The inclusion of schoolmaster in the list of Robert's activities throws some doubt on the song's authorship since the schoolmaster at Brassington was his kinsman Daniel, who taught the workhouse children and kept the accounts. Daniel was well placed to know what went on in the workhouse and literate enough to have written the song.

Thomas Westerne's and Job Marple's old inn, probably, by 1820, already the oldest house in the village, can never have been an ideal place to house large numbers of people. There was no water piped to the house and no well in the garden. Looking at the rooms now, it is impossible to imagine how seventy-seven men, women and children were fitted into them. This was the number recorded when the house was being used by the successor to the Brassington Incorporation, the Ashbourne Poor Law Union, between 1845 and 1848.[38] Earlier numbers seem to have been smaller, as might be expected when it was serving fewer parishes – the 1841 census listed the governor, William Torr and his wife, and fifteen inmates.

As well as buying the workhouse on behalf of the Incorporation, the village overseers seem to have run an early form of council-housing scheme. By the time of the 1835 poor rate survey, 'Brassington Township' owned eleven houses, seven of them with gardens and ten of them tenanted. When the Incorporation was wound up in 1844 its parishes were included in the new Ashbourne Union. In the opinion of the Assistant Poor Law Commissioner, Robert Weale, the Brassington workhouse was unsuitable – it was an 'ill arrangeable place', in fact. Plans were started for a new one in Ashbourne but in the meantime the Board of Guardians who administered the Ashbourne Union made do with Brassington. In addition to the seventy-seven places in the old Red Lion, they took sixty-three more by renting a second building from Philip Hubbersty of Wirksworth. This was the house a few yards away from the Red Lion which had been the George & Dragon and was to be so again later.

For three years the two old pubs were home to a small army of grey people: men and boys wearing grey suits and grey shirts, women and girls in grey gowns, grey petticoats and grey shifts. The men and boys were given black woollen hats and the women and

The George & Dragon in 1910, the 'upper house' of the Brassington workhouse from 1845-48.

girls coarse straw bonnets. Gilbert's Act had permitted well-behaved paupers to remove the capital P from their clothes, but the Ashbourne Guardians made sure that the Brassington villagers should not mistake the paupers for ordinary people: they ordered a supply of 'Union' buttons for them to fasten their clothes. The new regime was efficient. The room in the Red Lion facing the stone pit or quarry then in the grounds of the house was fitted out to increase the accommodation, William Knowles being paid £1 2s to make a new door and repair the floor.

A fence was built across the garden to make the quarry secure. This was where the able-bodied men of the workhouse spent their time, breaking one yard of limestone each day into pieces small enough to pass through a two and a half-inch ring. Two cast-iron boilers, of sixty gallons and thirty gallons respectively, were bought in Derby and installed in the Red Lion. Water was piped to the house from the second workhouse, higher up the street, which had a pump, and Mr Hubbersty was ordered to get the pump repaired. The workhouse schoolroom was in the 'upper house' and the children seem to have been moved there when the second house was taken over, as a minute records that, on Weale's recommendation, the former children's room should have a stove installed and be used as a laundry and drying room. A further refinement ordered by the Board was a wall to divide the privies, and the way to them, from each other.

Outdoor relief, dole, was available to paupers who had somewhere to live, as long as they arrived to collect it before 9 a.m. on 'Board day'. If they were caught begging the relief was stopped, and they were committed to the workhouse, where they were given food, clothing and shelter in return for servitude. The food was as adequate, and as dull, as the workhouse clothes. These temporary villagers breakfasted on eight ounces of bread and two pints of milk porridge, repeated for supper on every evening but Monday. Their midday dinner was nine ounces of meat, a pound and a half of potatoes, a pint of meat soup and eight ounces of bread on Sunday and Wednesday. These were the red-letter days. On Monday and Tuesday there were no meat and potatoes, on Tuesday and Friday dinner consisted of one pound of dumplings, and on Saturday the paupers had either Monday's soup and bread or Tuesday's dumplings. During the Irish potato famine of 1846 rice, peas and more bread took the place of the scarce potatoes.

The Board of Guardians was responsible for the health of the workhouse inmates and other paupers, and paid a physician, William Cantrell, £30 a year for attending to the paupers in the Brassington district and a further £25 for the workhouse patients. Efforts by Cantrell and other local medical officers to persuade the Board to increase their salaries were refused, in spite of Cantrell's success in getting the Poor Law Commissioners to rule in favour of increases. Cantrell made several attempts, in writing, to move the Guardians, perhaps the most telling being a letter asking them to supply four quarts of chloride of zinc during an epidemic among the paupers in 1847: 'I readily believe you have not a Guardian that would attend the Fever Cases for one Week, for my whole years salary, and Doctors have not charmed lives, as the death of one yesterday from Fever caught in discharge of Parochial duties shows.' Later in the year Cantrell put forward a plan for creating a fever ward and a vagrant ward in the upper workhouse, for which work the Guardians granted him £5.

The Guardians were also responsible for the education and religious instruction of the paupers. At their first meeting they resolved to ask the managers of Brassington School to admit the workhouse children. The managers were happy to allow the headmaster to teach the children, but only if they stayed in the workhouse, and there is a record of payments of £2 13s 2d and £2 0s 5d being made to Daniel Slack for teaching there in 1845. After Slack's death in 1846, it was decided to appoint a workhouse schoolmistress, and Ellen Unwin was appointed, at an annual salary of £17. Miss Unwin, far from her home in Chapel-en-le-Frith, found life hard in the workhouse. She fell foul of the matron, wrote a letter of complaint to the Board, was reprimanded by the Board for, allegedly, not keeping the children clean, and resigned in April 1847. The children's further education was at best the experience offered to young Ann Neal in 1847. The clerk to the Guardians applied 'to take her into his servitude on the following terms, The Girl to be clothed at the Union's expense, & to serve him till Christmas next for her food, he undertaking, if not re-engaged, to deliver her up with her Clothing equal to what it was, on entering his service.' This application for slave labour was granted.

The spiritual needs of the paupers were met by the Revd A. Weyman, an Anglican clergyman who had been appointed on a majority vote of the Guardians against a proposal that ministers of all denominations be allowed to attend the workhouse. The Board decided that dissenting ministers should only be allowed into the workhouse when inmates asked for them.

It was a hard regime for the staff there, as well as for the inmates. There is a minute of one meeting of the Board of Guardians at which James Banks, porter at Brassington, was given enough time off to go to church on Sundays. It is easy to imagine the Board's reply if Banks had put in a more secular request. The master, too, had a hard time of it. He was Samuel Walker of Ashbourne, appointed, with his wife as matron, in 1845. Walker's salary was £40 per annum and his wife's £25. For his £40 Walker operated at the sharp end of the Poor Law system and sometimes suffered from the frustration of its victims. In February 1848 the Board sent him a cheque for nineteen shillings to cover the expenses he had incurred in committing a vagrant to trial. The vagrant had attacked him. Walker was effective enough to keep his job when the new workhouse opened in 1848 and seems to have been humane too. For Christmas 1847 he was 'permitted to give the workhouse inmates a dinner of roast beef and plumb (sic) pudding on Christmas Day, as last year.'

The 'fever' which William Cantrell treated with zinc chloride was in fact a fatal epidemic of 'unparalleled sickness and malignant fever'. Mr Weale was asked to examine the new workhouse so that the paupers could be moved to Ashbourne as soon as possible. When they eventually made the journey to their fine new quarters, in July, their pleasure at being decently housed was probably equalled by the relief felt by the villagers at the removal of such a potent source of sickness.

The village had in fact benefited from the business which the workhouse had brought. This was greater during the days of the Brassington Incorporation, but even when the contracts were placed more widely by the Ashbourne Guardians there was still some work for the village farmers and tradesmen and women. The girls' straw bonnets were made by a Miss Davenport in Brassington, John Mellor supplied shoes at 7s a pair for men and boys and 4s 3d for the women and girls, and Thomas Slack tendered successfully in September 1845 to supply milk at seven quarts for 1s 1d, butter at 1s a pound and cheese at 6d a pound. Later in the year William Knowles carried out repairs to the workhouse, Thomas Allsop and William Fearn supplied stone and tiles and George Torr secured the hair-cutting and shaving contract. Brassington, however, was no place to house the paupers and vagrants of the parishes of a large Poor Law Union such as Ashbourne. For the three years they were there, the paupers made up about a fifth of the population of Brassington, and Robert Weale's judgment that the old inns made inadequate workhouses was proved to be a sound one.

VILLAGE TRAGEDIES

It was hard to make a living and there were those in the village who failed and ended up on parish relief or worse, in the house of industry. There were also the tragedies common to all communities at times like the suicide of Mrs Holloway in 1833, reported in *The Derby Mercury* of 26 June. When her husband discarded her she decided that life without him was not worth living and hanged herself: the coroner's jury decided that she was suffermg from 'temporary insanity'. Another suicide, reported on 17 September in the following year, seems to have happened because nineteenth-century doctors could do little to relieve pain. George Fearn, a stonemason, lived with his mother in the village. He went off to work one Friday morning and never came back. He was found on Sunday, hanging in his barn – 'the deceased had suffered very much from a rupture, with which he was afflicted, and it is supposed the pain drove him to a fit of insanity during which he suspended himself by a cord to the side tree in the barn.' On 25 May 1837 *The Mercury* reported the inquest on the newborn daughter of Mary Ann Heathcote, found dead at the bottom of the shaft in Fleelow mine. Mary Ann, a young servant, living away from her home in Biggin, was tried for the baby's murder and, unusually for those unforgiving days, was acquitted: *The Mercury* of 2 August recorded the testimony of several witnesses to her distress and bewilderment and the jury gave her the benefit of the doubt.

RELAXING

Most of the villagers, however, survived as they had always done and made the best of it. Their sense of community was shown in the building of the schools and the repairing of the church, and that 200 from a total population of about 700 should turn out for the meeting at Longcliffe in 1862 is a dramatic illustration of the importance of the chapels in their lives. They enjoyed the big occasions, when marching bands transformed the narrow streets, and they amused themselves in the old ways. Among the 'various kinds of sinful gratification' with which James Fearn passed his time before the Methodist preachers converted him was dancing, 'a sport in which he took great delight, and he indulged himself therein on every convenient occasion.' Fearn was especially fond of Wakes Week, spending every night in the pub 'with other young men and maidens, drinking, dancing, and singing the bacchanalian song.'

John Fearn's reminiscences at his golden wedding anniversary included memories of cock-fighting, bull-baiting and dog-fighting, all confirmed in Hackett's book in 1863[39]. Hackett contrasted the Wakes Weeks of his own day with those of thirty years earlier, when 'at these feasts the brutal sports of bull and bear-baiting and cock-fighting were resorted to.' He saw their disappearance as evidence 'of the moral advancement of the humbler classes', but admitted that Wakes Weeks were still, in 1863, likely to become unruly: 'brawls... do occasionally occur.' Wakes Week was still scandalising some chapel people at the end of the century. One local preacher's wife, in a letter to her daughter, deplored the antics inspired by drinking deceptive rhubarb wine, where two brothers fought in the street 'so you can see what good the wakes has been.'[40] Wakes Week was staged at the end of September in James

Fearn's time, and there was more to it than carousing and bull-baiting. A report in 1810 describes plays performed on temporary stages every night of Wakes Week in Brassington.[41]

The village band seems, by the beginning of the century, to have played a more familiar collection of instruments than the group which had played at the coronation of George III in 1761. This inference is drawn from the draft of a letter written in 1808 from the Miners Arms. It is in the book which the landlord had inherited from his father.[42] The letter is between the mining and pub accounts, and is clearly written by the bandmaster. He thanks his (unnamed) correspondent for sending him a march and comments that the band can play 'anything in reason for a country Band such a March as you sent first sight.' He goes on to contrast the leaders, or the men playing the melody instruments, with the ones playing the basses. The village had a band then, who could sight-read with a proper balance of instruments. Music was important. The landlord's accounts include debts incurred for fiddling at the celebrations which took place at the opening of a new mine, and for supplying ale to carol singers, as well as the bell ringers. The church and, later in the century, the three chapels, all had choirs, which performed as the United Choir on occasion, and there always seem to have been soloists prepared to stand up and sing at concerts.

Cruel sports like bull-baiting were clearly never the chief amusement of the 'humbler classes' in the village, and there is a surviving reminder of one of their gentler pastimes. On the hillside above the church is a small field, carved out of the slope and bordered by the rocks which its creators removed when it was made. This is called the Bowling Green and there is documentary evidence that the men of the village did indeed make their way up the hillside to play bowls there. In 1856 the manor-court recorded the sale of a number of fields in 'Upper Yearnstones'. Excluded from the sale was part of one of them 'which was formerly a bowling green'.[43]

REFERENCES

1 Farey, Vol. 12, 1813.

2 DRO D1297/P1.

3, 4, 6, 7, 20, 42, private collection.

5 DRO D2629z 2/1.

8 *Ashbourne News,* 2 April 1909.

9 MUJRL Methodist Archive.

10 DRO QS Dissent 1829.

11 DRO 1634j/MP 136.

12, 13, 14, 16, 18, 21, 26 & 27 WEA, 1967.

15 DRO D1983 J/C19.

17 White, 1857.

19 & 41 Unidentified magazine cutting.

22 Pigot, 1835.

23 NSEP, 1846.

24 DRO D250 c/W1/2.

25 DRO D1983 J/ C19.

27, 31, 33 & 42 DRO D2629 zl/l

28 Brassington School. School log.

29 DRO c/EAM1.

30 & 34 Farey, Vol. 3, 1817.

32 DRO D2629 Add 1.

35 DRO D166M book G.

36 DRO D239M.

37 DRO QAGp 288.

38 DRO D520 c/W1/1.

39 Hackett, 1863.

43 DRO D166M book K.

CHAPTER 10

TRANSFORMATION

TWENTIETH-CENTURY CHANGES

Photographs from the turn of the nineteenth century show a village which looks remarkably like the Brassington of 2007. Most of the houses shown in the old photographs are still here. There are some missing from Hillside and some from the south side of Church Street, near the Gate, but the main difference in the village is that the space between Town Street and Church Street, where the meadows used to be, has been filled by houses. Ashbourne Rural District Council built six between the two world wars and thirty-four after the Second World War, but Town Street, Miners Hill, Church Street, Maddock Lake and Kingshill would seem totally familiar to a nineteenth-century lead miner.

He would be disappointed, though, if he called at the Red Lion, Thorn Tree or George & Dragon for a drink after work, as only the Miners Arms and the Gate are still pubs. The miner would find the paved streets a great improvement on the wet or dusty streets he knew, and he would be profoundly grateful for two other twentieth-century improvements – electricity and tap water. The electric mains came to the village in 1930, street lamps a year later, and mains water in 1939.[1] In that year, the villagers ended their centuries-old routine of taking their buckets to the well. These innovations, followed in 1951 by the installation of a sewage scheme, were revolutions in the villagers' lives, making their everyday existences quite different from their ancestors'. The village looked the same, but there had been a bigger change in the experience of living there during a few years of the last century than during the whole of the previous three or four.

MINING PERSISTS

Some changes were delayed. There was still work underground for a few villagers until the 1950s. The mineral extracted from the old lead mines in the twentieth century was barytes, ignored until then but in demand as a source of barium. It was taken from Great Rake and Nickalum until 1919, from Conway Knowl until the 1940s, and from Golconda, the old Gell mine in Griffe, until 1953. Golconda, which was then about 160 years old and had become uneconomic as a lead mine, was bought by R.H. Key in 1915.[2] He modernised it to produce barytes, used in the steel industry and vital for the war effort. Compressed-air drills, tramways, a processing plant and a branch line to a siding on the High Peak Railway were installed. The mine employed about twenty men and one who worked there from 1926 until 1931 still lives in the village. Drills and waggons used in these latter days are still lying at the sides of the mine roads, over four hundred feet underground, at Golconda.

FARMING CHANGES

Until the 1950s there was little change in the economics or method of Brassington's beef and dairy farming and there are still sheep grazing the hilltop pastures of the former wastes of

Carsington Pastures and Brassington Moor. There were still farms in the village itself. *Kelly's Directory* of 1936 lists twelve farmers and four 'cowkeepers' living in the village and cattle continued to be driven through the streets to and from their milking sheds. Haymaking was partly mechanised. The grass was cut by horse- or tractor-drawn mowing machines and drawn into rows by horse-drawn rakes. There were machines which turned over the rows of mown grass – 'swathe turners' – and threw it about to air it ('tedders'). The hay was still, however, picked into carts, either drays or muck carts improved by 'gormers', or rails, to raise the sides, and led to stacks or barns. Even more ancient methods were still used. At least one of the 'cowkeepers', with a smallholding on the steep slopes of Yearnstones, cut the sparse grass with a scythe, turned it with a pitchfork, and then loaded the hay on to a tarpaulin and dragged it down to the stack. On all the farms the hay, by then compacted, was cut in winter with a broad-bladed knife, and fed to stalled milk cattle or to stirks wintering in the fields.

In 1936 there were thirty-one farmers in the whole parish, six of whom farmed over 150 acres. This was a drop of twenty-three from the 1881 census figure, and very many fewer men were needed to work their mowers, tedders, turners and horse-rakes than when lines of scythe-men cut the grass and the rows were turned and tedded by men and women with pitch forks. That there were still haybarns in the village and still cows driven along its streets and milked in sheds at West End, on Town Street and Nether Lane, meant that the sights, sounds and smells of farming still filled the village, but the farms employed only a minority of the people by the middle of the century. There has been fundamental change in the last seventy years.

Haymaking has become a task for one man, his tractor and a bailer, or has been superseded altogether by silage making. Machine milking has replaced the man or woman on the milking stool and dairy farming has moved out of the village to the larger farms in the parish. The only farming operated from the village itself is cattle- and sheep-grazing on the moors.

As farming followed mining into history, as far as being a large-scale employer is concerned, the decline in the village's population continued, to 532 in 1961. This was a fall of ninety-one from the 1951 figure and was the steepest drop in any decade for which there are records. It has been stable since then, falling slightly to 527 in 1971, rising to 558 in 1981 and falling again to 540 in 1991. By 2001 the village's population had risen to 584, bringing it well over the figure recorded forty years earlier.[3] Some of this increase came from a small housing development at the south of the village.

SELF-SUFFICIENCY

For some years Brassington, while declining as a working village, kept its old character. There was still a blacksmith at work in the village. In 1936, in addition to the farms, there were still a butcher and eleven other shops. Three of them sold clothes, including Brindley's, on Church Street, in the building which had been the George Inn in the eighteenth century. In 1936 Thomas Brindley was described in *Kelly's* as 'grocer, draper, clothier and patent medicine vendor'. Even more like a great universal store was Ernest Taylor's shop. He was a 'grocer, confectioner, tobacconist, ironmonger, wireless apparatus, cycle agent and battery charging'. In the village, Ernest Taylor's wirelesses would be plugged into the mains by 1936, but most of the outlying farmers would still need to bring their batteries in for charging. Another draper, Joseph Brown, sold and repaired shoes, and there were still two shoemakers: John Mellor on Kingshill and George Walker, carrying on his grandfather's old business. Stanley Allsop sold sweets, Frank Stevenson vegetables and fruit and Mrs Hall carried on a small trade in drapery from her home at Stile House. There were a newsagent, Mrs Yates, and a branch of the Wirksworth and District Co-op. At Maddock Lake, in the middle of the village, Oulsnam and Fearn's 'steam saw mill and timber yard' added the sounds and smells of sawing wood to the village's working atmosphere.

Waiting at the well, as their ancestors had done for centuries: a scene that vanished forever when water was piped to the village in 1939. Note the now-demolished houses in the left background.

Making hay the old, hard way.

Dick and Joshua Taylor outside their smithy at Dale End, about 1900.

An early twentieth-century football team outside the Miners Arms. The landlord, Alfred Charlton, stands at the right, in his best suit, rosette and bowler hat.

Brassington Band poses during the 1920s.

In 1936 the men of the village had two main jobs to chose from, in addition to farming. There was the Swan, Ratcliffe brickworks at Hopton, and quarries at Hoe Grange, Longcliffe and Grange Mill, all served by the Cromford and High Peak Railway. By 1962[4] there were still twenty-three working in the quarries and sixteen at the brickworks, though much of the railway's trade had been transferred to road: nineteen of the villagers were lorry drivers. There were still forty-eight working on farms in 1962. By 1980 there were twenty-four working the farms, about thirty lorry drivers and twenty quarrymen.

The brickworks had closed in 1971, the railway in 1976. In place of Oulsnam's timber yard was Robinson's steel fabrication works, employing twelve people. Robinson's is still there and close to it, in the centre of the village, is a pine-furniture workshop, making furniture to customers' own specifications. At the West End, a haulage business is run from Brookfield, the former Bucksleather House and there are two mineral processing plants near the village. However, the 2001 census found that 71.9 per cent of Brassington's workers travelled to work by car and another 1.1 per cent by bus or taxi.

SOCIAL LIFE

While the post-medieval pattern lasted, while there was work in or near the village, while there were still men farming a few acres and while most of the necessities of life could still be bought in village shops, many of the old institutions survived. For more than half of the twentieth century, Brassington had cricket and football teams, a brass band, three chapels, the Oddfellows and the Royal Antediluvian Order of Buffaloes. After the shock of the war of 1914-1918, during which Brassington suffered along with the rest of the country, with fifteen of its men killed, the villagers resumed a vigorous social life. For forty years, between 1919 and 1959, the Brassington Reading and Recreation Society met in the 1832 school building on Hillside, run by an elected committee, usually chaired by the vicar. There was a secretary and a treasurer and proper minutes were kept of the committee's decisions.[5]

Subscriptions were 2s 6d a quarter, soon rising to 3s. At the committee's second meeting, on 27 September 1919, it was resolved that a tender of £2 10s for cleaning the room be accepted from 'the woman that brings along the sticks to lay the fires', and that the secretary should see Mr Oulsnam about sweeping the chimneys. Twelve men were elected to serve as attendants, in pairs, on weekdays and nights 'and to take charge of the billiard table'. The names of the committee-men and others were the old village names from the nineteenth and earlier centuries – Watson, Hall, Taylor, Hodgkinson, Allsop, Fearn, Gerrard, Brindley, Swindells, Seals, Radford, Heathcote, Buckley. The room was only opened when two members, including a member of the committee, were there to take charge. The society ran a billiards handicap, with a first prize of £1 and a second of 10s. The committee-men were authorised to order any member who had been 'unruly or disrespectful to any member' to leave the room, and to suspend him until he had appeared before the committee to make his peace. The subscriptions paid for repairs, decoration and glazing 'the windows in the top room'. There were times when the room was closed for lack of funds, but they were few and this was a strong and popular social club for the men of the village.

There were dances and whist drives in the school, which had the large room divided by a hinged screen which was common in nineteenth-century schools. With the screen pulled back the village had a hall where they could dance to the music of the Tudor Band, the Windsor Band, John Spencer's band, Tim Wray's band. In a typical year there were dances on Easter Monday, 23 April, Whit Monday, 18 June, 16 July, 2 August, on 1 September and the second in Wakes Week, and on 27 December and New Year's Eve. These events, and the annual carnival and Wakes Week, were being organised after the

Second World War by the Village Hall Committee, with the intention of raising money to buy enough land to build a village hall; in 1948 they had reached agreement with Ashbourne Council to buy a piece of ground at the north of the council estate for £100. The village in fact had to wait for its hall until 1982.[7] The post-war effort raised about £1,000, and a new committee, formed in 1972, had raised another £5,000 in time to profit from the closure of the Congregational chapel in 1977. With the help of a local authority grant the old chapel was transmogrified and became the village hall five years later. However, it has done the job for twenty-five years and the village is now, for the second time, in the process of planning and raising money for a purpose-built hall to replace it, this time at the west end of the village.

The Congregational chapel's failure in 1977 was followed by the closure of the Primitive Methodist chapel in 1985. Both had flourished for the first half of the century. Until the 1950s the Primitive Methodists had a Sunday school with about twenty children on the register and four regular teachers.[8] For the whole of the inter-war period the collection at the Sunday evening service was around 15s, rising to £1 during the war, implying a regular congregation of twenty to thirty. By the time the Hillside chapel celebrated its 150th anniversary in 1984, the congregation had almost disappeared, the evangelical force of Methodism spent. A similar decline affected the third chapel in the village, the Wesleyan Reform at West End, and this too is now closed. The decline in chapel-going has been part of a general retreat of organised religion from the centre of the villagers' lives and the older church, too, is much contracted from the days when it was vigorous enough for Brassington to be made a separate parish after centuries of being part of Bradbourne. The villagers still get married in the old church, still take their babies to be christened there and, until recently, were still buried in the churchyard – a new burial ground was opened at West End twenty years ago. The congregation, however, is too small to require the undivided attention of the vicar, who ministers to Bradbourne and Ballidon, as well as Brassington.

The 1950s saw the end of cricket and brass banding. Both had flourished in the nineteenth century but in the second half of the century interest seems to have waned, and the village failed to produce its own musicians and games players. The cricket team, playing on barely suitable pitches at Wash Hills, Harborough and Longcliffe, had existed at least as early as 1862. A newspaper report on a match played between Brassington and Alderwasley in that year is reprinted in the 1967 village history[9], which goes on to give the results of matches played between 1919 and 1938. The team's scores were usually low, especially on their own rough pitches, making the 114 for 5 reached in 1927 against Youlgreave a triumph – the wicket cannot have been any better than usual as Youlgreave managed only 5 runs.

That the village had good players in the inter-war years is apparent from such scores as the 129 on the Rolls Royce ground in 1934 (W. Brindley 59, E. Brittain 52 not out). Ted Brittain was an all-rounder at cricket (5 for 15 against Mayfield in 1939), a fine footballer, and gave a start of twenty points to the next-best player in the billiard handicaps. The cricket club was more than a team of cricketers. They organised the carnival in 1935 and made a profit of £30 9s 6d, 75 per cent of which was given to the Derbyshire Royal Infirmary.[10] This carnival included a dance at which the music was played by the Night Hawks Band at a fee of £2 15s and which raised £20 3s 6d, plus £5 4s 5d from the sale of refreshments – most of the village must have been there.

There are photographs of the cricket teams and of several twentieth-century football sides. The footballers played on fields at Bradbourne Lane, Kilcroft, near the Hall, the Green and at Wash Hills. They won the Ashbourne Cottage Hospital Medals by beating Ashbourne Town reserves in 1900, and the history lists a string of local honours up to 1953. The team won the Cavendish Cup in 1952, though it has to be said that the village players had some help from local stars brought into the side then and in the following

The Brassington Middleton and Wirksworth Band leads the 1989 Wakes procession into Miners Hill, with Tudor House in the background.

year, when they won the Derbyshire Medals competition. That was the last season for a good many years, but the village team has in recent years been resurrected, and plays competitive football in a local league. There are plans for a new pitch.

Mid-century was also the time when the village lost its band. It had had one at least since the celebrations for George III's coronation in 1761, and probably for very much longer. Inter-war photographs show the bandsmen, sometimes in uniform with peaked caps, sometimes not, leading parades through the village. They played at all the outdoor events. The village hall committee's accounts, for instance, show a payment to 'Brassington Silver Band' of £3 10s for playing at the Wakes in August 1946. They appear regularly thereafter for similar fees: by 1950 it had risen to £5 for the Wakes parade. The village was losing interest in its band, however, and, like the chapels, the church and the cricket and football teams, it was short of bodies. It found it hard to find enough players to lead a parade or take on a concert, and in 1964 it amalgamated with the Wirksworth and Middleton bands to form the BMW Band, practicing at Wirksworth.

The Brassington Band spirit lingered a few years – in 1968 one of its old players won a court case brought against him by the new band, who claimed that he had a cornet and two uniforms which should have been handed over in 1964.[11] He insisted that the Brassington Band still existed and when the BMW Band's solicitor told him that there was in fact no band in Brassington and that he did not practice his cornet playing, he replied, 'I do practice. I sit at home and play the organ with one hand and the cornet with the other.' Although agreeing that he did not wear his uniform, he buttressed his unlikely tale of two-handed music-making by claiming that the band still had trustees, a secretary and playing members. It was the BMW Band, however, which led the parade at Wakes Week. There is still a Wakes Week in the last week of July, and though there is no longer a procession through the village, a concert by the BMW Band is one of the events staged.

Rapid rises in house prices throughout the country in the early 1970s, in the mid-1980s, and in the last few years, coupled with a fashion for country living which has pushed up the prices of village houses to levels which put them outside the reach of most villagers, have produced the same effects in Brassington as in most country places. The village has its share of weekend cottages and most of its people now work outside it – there is a truly rural calm during weekdays. The working village of 1881 changed slowly but change it did. As the jobs went, so did many of the families who depended on them, and this smaller village is no longer the self-sufficient community it was up to forty years ago. It now has no shops, no post office, no petrol station.

Brassington has lost much, and it would not be surprising to find that its community spirit had suffered. However it still has its school, vital to any sense of community, and its two pubs. One of these, the Gate, runs a boules club, presumably started by after a visit to France by the landlord or one of his customers. The village has thriving institutions. A branch of the British Legion has flourished since 1932, meeting until recently in a wooden hut off Church Street, and since 1994 in a larger one. The new hut was given a limestone cladding in 1996. The Women's Institute, which currently has about thirty-two members, who meet in the village hall, celebrated its seventy-fifth anniversary in 2007. There is a Brownie pack, art classes, a parent and toddler group, an over-sixties' club, a ladies' club and a gardening club. During Wakes Week, there is an 'Old Uns versus Young Uns' football match, an open-air service on Miners Hill, a quiz night, boules at the Gate, a car treasure hunt and a village jumble trail. And there is a village website – brassington.org – from which most of the information in this paragraph was taken. Clearly still a community.

Of all the transformations through which the village has passed in its 1,400 years, those of the twentieth and twenty-first centuries have been the most complete. In changing from a place of work to one which is valued as a pleasant place to visit or to settle in, Brassington has simply changed with the times, as it has done since the Anglians moved in and founded Branzingtune.

REFERENCES

1, 4 & 9 WEA, 1967.

2 Slack, 2003.

3 Derbyshire County Council, 2007.

5, 6, 8 & 10, private collections.

7 *Ashbourne News Telegraph*, 6 May, 1982.

11 *Derby Evening Telegraph*, 23 April, 1968.

BIBLIOGRAPHY

Abbreviations
CHBC Chatsworth House, Barmaster's Collection
DAJ Derbyshire Archaeological Journal
DAS Derbyshire Archaeological Society
DLSL Local Studies Library, Derby
DNB *Dictionary of National Biography*
DRO Derbyshire Record Office
DRS Derbyshire Record Society
LJRO Lichfield Joint Record Office
MLSL Local Studies Library, Matlock
MUJRL Manchester University John Rylands Library
NSEP National Society for the Education of the Poor
PDMS Peak District Mines Historical Society
NA National Archive, Kew, London
SCA Sheffield City Archive
VCH Victoria County History
WEA Workers' Educational Association

Archives
Chatsworth House
   Barmaster's Collection
Derby Local Studies Library
   Gale Bequest, bundle 3

Newpapers
Derbyshire Record Office
   D166M, books A-M. Brassington manor court, 1635-1888
   D166M, boxes A166, A167, A168. Brassington manor-court papers, eighteenth and nineteenth centuries
   D258 Gell collection
   QS Quarter Sessions
   D161B Stone and Symonds collection removal orders
   D1297/P1 Brassington town map, *c.* 1820
   D2629 z2/1 Survey valuation and rate of the township of Brassington, 1835 (photocopy)
   D1634 J/MP 136 Primitive Methodist Winster Circuit membership lists, 1833-38
   D250c/Wl/l, 2 Ashbourne Poor Law Union minute book, 1845-47, 1847-48
   C/EAM 1 Brassington School Board minute book
   D239 Fitzherbert collection
   Q/RIc5 Brassington Inclosure Award, 1808
   Q/RIc6 Brassington Inclosure map, 1808
   D2629 z 1/1 Brassington measuring book, 1792 (photocopy)
   Brassington measuring book, 1820 (photocopy)
   D2629z 2/1 Brassington Poor Rate and Land Tax Survey, 1835 (photocopy)
   D2629 (Add 1) Brassington Oddfellows contributions book (photocopy)
   Brassington parish registers, 1716 onwards
Lichfield Joint Record Office
   Probate B/C/11 wills and inventories
   B/C/5, 1636, 1637 Brassington tithes
   B/A3 Brassington presentations, 1727
   B/V/5 Primary visitation, 1772-
   Brassington bishop's transcripts, 1673

Matlock Local Studies Library
  British Museum added MSS (Wolley MSS) (microfilm)
  Census returns (microfilm)
  Lambeth Palace Library MSS 697 f137
  Derbyshire poll book, 1734
Manchester University John Rylands Library
  Methodist archive
National Archives, London
  DL30/43/484-DL30/54/674 Court rolls, Wirksworth wapentake and hundred, including Brassington manor
  DL43/1/19 Brassington rental, 1620
  DL43/20/2 Brassington rental, 1620
  DL29/369/6180 Honour of Tutbury accounts, 1441
  DL28/26/2 Honour of Tutbury accounts, 1505-42
  E317 Derb 9 Survey of the rents... of Wirksworth, 1651
  E317 Derb 28 Survey of the soke and manor of Wirksworth, 1649
  E101 59/7 Muster roll, 1538
  SP 12/3 (1558), 12/117 (1577), 12/139 (1580), 16/405 (1638) Muster rolls
  E101/65/19 Muster rolls, 1602
  E179/245/7 Hearth tax returns, Derbyshire, 1662
  E179/94/402 Hearth tax returns, Wirksworth hundred, 1664
  E178/245/9 (roll 3) Hearth tax returns, Wirksworth hundred, 1672
  DL30/54/669A Wirksworth Barmote Court, 1630
  E101/280/18 Petition of Derbyshire miners to the House of Commons for relief from the tax on lead, 1642-43
Sheffield City Archives
  S112 Compotus roll of Henry Stafford, Receiver of John, Earl of Shrewsbury... 1446-47 (translation in Hunter Archaeological Society Transactions, vol. 2. pp 344-59)
  A128 Rentals for the Earl of Kent's manors (microfilm)
  Bacon/Frank collection 2/44, 2/148
  ACM S116 A book of estreates, paynes and presentments, 1578
  ACM S114 Rentals, 1581-84
  ACM S117 Rentals, 1587-92
  ACM S118 Rentals, 1587-93
  WD924, 1596
  Talbot correspondance, 2/78, 2/79

Printed works

Anderson, J.J. *Roman Derbyshire*. Hall, 1985
Beckett, J. *The Agricultural Revolution*. Blackwell, 1990
Bennett, H.S. *Life on the English Manor*. CUP, 1937
Bill, E.G.W. *Calendar of the Shrewsbury Papers in the Lambeth Palace Library*. DAS, 1965
Birrell, J.R. *The Honour of Tutbury in the Fourteenth and Fifteenth Centuries*. Birmingham University thesis, 1962
Blair, P.H. *An Introduction to Anglo-Saxon England*. 2nd ed. CUP, 1977
Blanchard, I.S.W. *The Duchy of Lancaster's Estates in Derbyshire, 1485-1540*. DAS, 1971
Blanchard, I. S. W. *Economic Changes in Derbyshire in the Late Middle Ages, 1272-1540*. London University thesis, 1967
Branigan, K. 'Two Roman Lead Pigs From Carsington'. *DAJ*, 106, 1986, 5-17
Brooks, N. 'A New Charter of King Edgar'. *Anglo-Saxon England*, 13, 1984, 137-155
Cameron, K. *The Place-Names of Derbyshire*. 3 vols, CUP, 1958
Carrington, W.A. *Papers Relating To Derbyshire Minsters Temp Queen Elizabeth*, DJA, 17, 1895
Childs, J. *A History of Derbyshire*. Phillimore, 1987
Clark, P. *The English Alehouse*. Longman, 1983
Clark, R. 'Lists of Derbyshire clergymen, 1558-1662'. *DAJ*, 104, 1984, 19
Cox, J.C. 'An Elizabethan Clergy List of the Diocese of Lichfield'. *DAJ*, 6, 1884, 157
Cox, J.C. 'Minute book of the Wirksworth classis, 1651-1658'. *DAJ*, 2, 1880, 135-222
Cox, J.C. *Notes on the Churches of Derbyshire*. Vol. II, 1877
Cox, J.C. 'A Religious Census of Derbyshire'. *DAJ*, 7, 1885, 31-36
Currey, P.H. 'The Saxon Window in Mugginton church'. *DAJ*, 25, 1903, 225
Defoe, *A Tour Through the Whole Island of Great Britain*. Dent, 1975
Derbyshire County Council. Census 2001. Brassington profile
Derbyshire County Council. Listed buildings.
Dictionary of National Biography

Dodd, A.E. *Peakland Roads and Trackways*. 3rd ed. Landmark, 2000

Dool, J. 'Roman Material from Rainster Rocks, Brassington'. *DAJ* , 96, 1976, 17–22

Dugdale, W. *The Visitation of Derbyshire, 1662-1664*. Harleian Society, 1989

Edwards, D.G. (1) 'Derbyshire Hearth Tax Assessments, 1662-70'. *DRS*, 1982

Edwards, D.G. (2) 'Population in Derbyshire in the Reign of Charles II'. *DAJ*, 102, 1982, 106–117

Ernle, Lord. *English Farming Past and Present*. 6th ed. Heinemann, 1961

Farey, J. *A General View of the Agriculture and Minerals of Derbyshire*. 3 vols 1811–17

Ford, T.D. and Rieuwerts, J.H. *Lead Mining in the Peak District*. 4th ed. Landmark, 2000

Glover, S. *History, Gazetteer and Directory of the County of Derby*. Part II, 1829

Hackett, R.R. *Wirksworth & Five Miles Around*. 1863

Hardy, W. *The Miner's Guide*. 2nd ed. 1762

Hart, C.R. *North Derbyshire Archaeological Survey*. 1981

Hey, D. *Packmen, Carriers and Packhorse Roads*. Leicester University Press, 1980

Hodges, R. 'Roman or Native in the White Peak'. *DAJ*, 101, 1981, 42–57

Hoskins, W.G. *The Midland Peasant*. Macmillan, 1957

Hunt, W.H. 'A list of the "Alehouses, Innes and Tavernes" in Derbyshire in the year 1577'. *DAJ*, 1, 1879, 68–80

Ince, T.N. 'Wirksworth, Bonsall, Brassington, and Ireton Wood, certain copyholds confirmed'. *Reliquary*, 12, 1871-2,253

Jeayes, I.H. *Descriptive Catalogue of Derbyshire Churches*. 1906

Kerry, C. 'Darley Abbey Charters Preserved at Belvoir'. *DAJ*, 16, 1894, 14–43

Kiernan, D. 'The Derbyshire lead industry in the 16th century'. *DRS*, 1989

Langdon, J. 'Horses, oxen and technological innovation'. CUP, 1986

Ling, R. 'Excavations at Carsington, 1978-80'. *DAJ*, 101, 1981, 58–87

Ling, R. 'Excavations at Carsington, 1983-84'. *DAJ*, 110, 1990. 30–55

Marsden, B.M. *The Burial Mounds of Derbyshire*. 1977

Milward, R. *A Glossary of Household, Farming and Trade Terms From Probate Inventories*. 2nd ed. DRS, 1982

National Society for the Education of the Poor Survey, 1846

Pevsner, N. *The Buildings of England: Derbyshire*. 2nd ed. Penguin, 1978

Pigot, J. & Co. *National Commercial Directory for Derbyshire*. 1835

Richardson, J. *The Local Historian's Encyclopedia*. 2nd ed. Historical Publications, 1986

Riden, P. 'The Population of Derbyshire in 1563'. *DAJ*, 98,1978,61–71

Saltman, A. 'Cartulary of Dale Abbey'. *DAS,* 1966

Slack, R. 'A Survey of Lead Mining in Wirksworth wapentake, 1650'. *PDMHS Bulletin*, 10, 4, 1988, pp. 213–216

Slack, R. 'The Dovegang Plot'. *PDMHS Bulletin*, 12, 3, Summer 1994, pp. 102–107

Slack, R. *Man at War: John Gell in his Troubled Time*. 1997

Slack, R. 'The Miner's Tale: Golconda, 1916-1957'. *Mining History*, 15, 3, Summer 2003

Stafford, P. *The East Midlands in the Early Middle Ages*. Leicester University Press, 1985

Stokes, A.H. *Lead and Lead Miners in Derbyshire*. PDMHS, 1964 (first published in 1880)

Victoria County History of Derbyshire.Vol. 2, 1907

Walcott, M.E.C. 'Church Goods and Chantries of Derbyshire in the XVIth Century'. Reliquary, 11, 1870-71

White, F. *History, Gazetteer and Directory of the County of Derby*. 1857

Winchester, A. *Discovering parish boundaries*. Shire, 1980

Wood, A. *The Politics of Social Conflict*. CUP, 1999

Wood, A. 'Beyond Post-Revisionism?' *Historical Journal*, 40, 1, 1997, pp. 23–40

W.E.A. *Life and History of Brassington Village, Derbyshire*, 1967

W.E.A. *Life and History of Brassington Village, Derbyshire: Supplement 1*. 1971

Yeatman, J.P. *The Feudal History of the County of Derby*. Vol. II, 1886